# Environment and Social Theory

Thinking about the environment, its meaning, significance and value is as old as human society itself. Yet it is clear that the present human generation is faced with a series of unique environmental dilemmas, largely unprecedented in human history.

*Environment and Social Theory* introduces the ways in which the environment has been used and abused, constructed and contested within social theory. Beginning with an overview of the place of environment within the history of social theory, this book examines the value and role of environment in 'classical' nineteenth century theory. Moving on to the twentieth century, Barry examines: gender, social theory and the environment; contemporary social theory and the environment: 'risk society' and postmodern approaches to the environment; biology, ecology and social theory; and concludes with a discussion on the possibility of a 'green' social theory.

Examining the ideas and views of key theorists including Hobbes, Locke, Freud, Habermas, Giddens and Beck, this book presents a comprehensive introduction to social theory and the environment.

**John Barry** is a lecturer in Politics at Keele University

# Routledge Introductions to Environment Series
## Published and Forthcoming Titles

*Titles under Series Editors:*
**Rita Gardner and A.M. Mannion**

### Environmental Science texts

Environmental Biology
Environmental Chemistry and Physics
Environmental Geology
Environmental Engineering
Environmental Archaeology
Atmospheric Systems
Hydrological Systems
Oceanic Systems
Coastal Systems
Fluvial Systems
Soil Systems
Glacial Systems
Ecosystems
Landscape Systems
Natural Environmental Change

*Titles under Series Editors:*
**David Pepper and Phil O'Keefe**

### Environment and Society texts

Environment and Economics
Environment and Politics
Environment and Law
Environment and Philosophy
Environment and Planning
Environment and Social Theory
Environment and Political Theory
Business and Environment

### Key Environmental Topics texts

Biodiversity and Conservation
Environmental Hazards
Natural Environmental Change
Environmental Monitoring
Climatic Change
Land Use and Abuse
Water Resources
Pollution
Waste and the Environment
Energy Resources
Agriculture
Wetland Environments
Energy, Society and Environment

Environmental Sustainability
Gender and Environment
Environment and Society
Tourism and Environment
Environmental Management
Environmental Values
Representations of the Environment
Environment and Health
Environmental Movements
History of Environmental Ideas
Environment and Technology
Environment and the City
Case Studies for Environmental Studies

**Routledge Introductions to Environment**

# Environment and Social Theory

John Barry

London and New York

First published 1999 by Routledge
11 New Fetter Lane, London EC4P 4EE

Simultaneously published in the USA and Canada
by Routledge
29 West 35th Street, New York, NY 10001

*Routledge is an imprint of the Taylor & Francis Group*

Typeset in Times by Keystroke, Jacaranda Lodge, Wolverhampton
Printed and bound in Great Britain by TJ International Ltd, Padstow, Cornwall

*British Library Cataloguing in Publication Data*
A catalogue record for this book is available from the British Library

*Library of Congress Cataloging in Publication Data*
Barry, John,
        Environment and social theory / John Barry.
            p.  cm. — (Routledge introductions to environment series)
    (Environment and society)
        Includes bibliographical references and index.
        1. Human ecology—Philosophy.  2. Human ecology—Religious
    aspects.  3. Social sciences—Philosophy.  4. Human beings—Effect
    of environment on.  I. Title.  II. Series.  III. Series:
    Environment and society.
    GF2.B28  1999
    304.2—dc21    99–10360

ISBN 0–415–17269–1 (hbk)
ISBN 0–415–17270–5 (pbk)

# Contents

# Series editors' preface
# *Environment and*
# *Society titles*

The 1970s and early 1980s constituted a period of intense academic and popular interest in processes of environmental degradation: global, regional and local. However, it soon became increasingly clear that reversing such degradation would not be a purely technical and managerial matter. All the technical knowledge in the world does not necessarily lead societies to change environmentally damaging behaviour. Hence a critical understanding of socio-economic, political and cultural processes and structures has become, it is acknowledged, of central importance in approaching environmental problems. Over the past two decades in particular there has been a mushrooming of research and scholarship on the relationships between social sciences and humanities on the one hand and processes of environmental change on the other. This has lately been reflected in a proliferation of associated courses at undergraduate level.

At the same time, changes in higher education in Europe, which match earlier changes in America, Australasia and elsewhere, mean that an increasing number of such courses are being taught and studied within a framework offering maximum flexibility in the typical undergraduate programme: 'modular' courses or their equivalent.

The volumes in this series will mirror these changes. They will provide short, topic-centred texts on environmentally relevant areas, mainly within social sciences and humanities. They will reflect the fact that students will approach their subject matter from a great variety of different disciplinary backgrounds; not just within social sciences and humanities, but from physical and natural sciences too. And those students may not be familiar with the background to the topic, they may or may not be going on to develop their interest in it, and they cannot automatically be thought of as being at 'first-year level', or second- or third-year: they might need to study the topic in any year of their course.

The authors and editors of this series are mainly established teachers in higher education. Finding that more traditional integrated environmental studies or specialised academic texts do not meet their requirements, they have increasingly met the new challenges caused by structural changes in education by writing their own course materials for their own students. These volumes represent, in modified form which all students can now share, the fruits of their labours.

To achieve the right mix of flexibility, depth and breadth, the volumes, like most modular courses themselves, are designed carefully to create maximum accessibility to readers from a variety of backgrounds. Each leads into its topic by giving adequate introduction, and each 'leads out' by pointing towards complexities and areas for further development and study. Indeed, much of the integrity and distinctiveness of the Environment and Society titles in the series will come through adopting a characteristic, though not inflexible, structure to the volumes. Each introduces the student to the real-world context of the topic, and the basic concepts and controversies in social science/humanities which are most relevant. The core of each volume explores the main issues. Data, case studies, overview diagrams, summary charts and self-check questions and exercises are some of the pedagogic devices that will be found. The last part of each volume will normally show how the themes and issues presented may become more complicated, presenting cognate issues and concepts needing to be explored to gain deeper understanding. Annotated reading lists are important here.

We hope that these concise volumes will provide sufficient depth to maintain the interest of students with relevant backgrounds, and also sketch basic concepts and map out the ground in a stimulating way for students to whom the whole area is new.

The Environment and Society titles in the series complement the Environmental Science titles which deal with natural science-based topics. Together this comprehensive range of volumes which make up the Routledge Introductions to Environment Series will provide modular and other students with an unparalleled range of perspectives on environmental issues, cross referencing where appropriate.

The main target readership is introductory level undergraduate students predominantly taking programmes of environmental modules. But we hope that the whole audience will be much wider, perhaps including second- and third-year undergraduates from many disciplines within the social sciences, science/technology and humanities, who might be taking occasional environmental courses. We also hope that sixth-form teachers and the wider public will use these volumes when they feel the need to obtain quick introductory coverage of the subject we present.

David Pepper and Phil O'Keefe
1997

## Series International Advisory Board

Australasia: Dr P. Curson and Dr P. Mitchell, Macquarie University

North America: Professor L. Lewis, Clark University; Professor L. Rubinoff, Trent University

Europe: Professor P. Glasbergen, University of Utrecht; Professor van Dam-Mieras, Open University, The Netherlands

 **Figures**

# Boxes

# Acknowledgements

I dedicate this book to my daughter Saoirse whose arrival midway through its completion both acted as a spur to finish it and at the same time provided a delightful distraction, which countered the latter and explains why it took longer than expected! I hope that by the time she is old enough to read it, the natural world will be in a better shape, or at least in no worse shape, than the sorry state it's in now. To my wife, Yvonne, I owe a great deal, not just for her love and support but for her many prescient observations and suggestions.

I would also like to thank my sister Deborah Barry who, as someone who was completely new to this area, acted as an invaluable 'guinea pig' in reading and commenting on the manuscript. The book is better for her suggestions as to how to make it clearer and more reader-friendly. Mark Charlesworth read and made comments on Chapters 2 and 3, for which I would like to thank him, while my colleague Paula Casal read Chapter 8, for which I am grateful.

Some of the material in the book has been based on various under-graduate and postgraduate modules on environmental moral and political theory here at the Department of Politics at Keele University. I thank all those students who took those modules and from whom I learnt a great deal.

David Pepper, the series editor, also deserves a word of thanks, not only for initially contacting me asking if I would be interested in writing this book, but for reading the manuscript and making many well-taken observations and helpful suggestions. I would also like to thank the anonymous referees for their comments, which were in the main extremely helpful.

Finally I would like to thank the long-suffering production team at Routledge (particularly Casey Mein) who were extremely understanding of my many missed deadlines.

John Barry,
Keele, November 1998

# Introduction: the environment and social theory

> The 'control of nature' is a phrase conceived in arrogance, born of the
> Neanderthal age of biology and philosophy, when it was supposed that
> nature existed for the convenience of man.
>
> (Carson, 1962: 100)

While it may be true that there is nothing new under the sun, when it
comes to discussing environmental issues it seems that perhaps this is not
the case. The novelty of environmental issues and problems, from global
warming, climate change and biodiversity loss, to concerns for animal
rights or the intrinsic value of nature, should not blind us to the fact that
humans have always thought about their relationship to the environment.
As such the environment and our relationship to it is a long-established
issue in social theory.

The newness of environmental concerns is more apparent than real in
that thinking about the environment, its meaning, significance and value
is as old as human society itself. However, it is clear that the present
human generation is faced with a series of unique environmental
dilemmas, largely unprecedented in human history. The present human
generation is the first one, for example, to have the capacity to destroy
the planet
many times over, while at the same time, it is also the first generation for
whom the natural environment cannot be taken for granted. So while
the environment has been a perennial theme in human thought, the
environment and how humans value, use and think about it has become
an increasingly central and important aspect of recent social theory and
political practice.

The overall aims of this book are the following:

- to introduce and discuss the ways in which the environment has been
  used and abused in social theory both past and present;
- to trace some of the historical origins of this relationship and to

demonstrate the importance of the environment for social theory;
- to examine key concepts such as 'environment', 'human', 'nonhuman' and 'nature' and related concepts in social theory;
- to explore some of the ideological uses of the environment in social, political and moral thought;
- to outline how some central thinkers and forms of social theory have thought about the environment;
- to outline both the 'greening' of recent social theory and the development of a green social theory.

## Outline of the book

The book is roughly divided into two parts. The first, historical, part (Chapters 2–4), offers a chronological discussion of past and present uses (and abuses) of the environment in social theory, from Judeo-Christianity to contemporary social theory. The second part (Chapters 5–8) looks at a variety of contemporary social theories, from economics, to gender, postmodernism, risk society and recent attempts to integrate ecological and biological thinking into social theory.

Chapter 1 defines how social theory is understood and used in the book, and looks at some general conceptual issues of social theorising and the environment. These general introductory issues include how we define what we mean by 'environment', and how this is related to such terms as 'nature' and 'nonhuman'. This chapter also looks at how and why the environment is used and abused in social theory, particularly when environment is understood as 'nature' or the 'natural environment'. Four dominant models or understandings of the environment which are often used in social theory are then outlined, and the chapter ends with a discussion of one of the most common ways in which the environment is used in social theory in terms of 'reading-off' principles or how society should be from observations of how the natural world operates.

Chapter 2 outlines the historical relationship between social theory and the environment. It focuses on exploring the Judeo-Christian legacy and moves on to then examine the Enlightenment as a key turning point in social theorising about the environment. More specifically, it uses reactions to the industrial and democratic revolutions to organise the discussions of how the place, role and power of environment varied in different forms of social theory.

Chapter 3 looks at pre-Enlightenment and Enlightenment social theorising about the environment. After a brief discussion of the different views of classical political philosophers such as Hobbes, Locke and Rousseau, this chapter then looks at nineteenth-century social theory and the environment. The centrality of social theory being 'scientific' is examined before moving on to analyse progressive and reactionary social theorising about the environment. The latter proceeds by focusing on four social theorists, Malthus, Darwin, Spencer and Kropotkin. Following on from a discussion of the classical Marxist analysis of the relationship between society and the environment, is an examination of the liberal perspective, focusing on the work of John Stuart Mill.

Following on from the previous chapter, twentieth-century social theory and the environment is the topic of Chapter 4. It begins with classical sociology, then moves on to a discussion of some key twentieth-century forms of social theory, such as the work of Sigmund Freud and existentialism. It then discusses the critical theory of the Frankfurt School of social thought and its critique of the 'dark side' of the Enlightenment and modern societies. This section focuses on how the domination of the external environment by modern societies (via the application of science and technology) can lead to the domination and distortion of human social relations and internal human nature. This then leads into a discussion of Herbert Marcuse and his take on the relationship between the domination of the natural environment and human emancipation. The latter half of the chapter is taken up with a critical examination of some recent social theory and the environment, focusing on the work of Jürgen Habermas and Anthony Giddens respectively.

The issue of gender is a key issue within modern social theory, and Chapter 5 looks at the insights and necessity of adopting a gendered analysis of the relationship between society and environment. It begins with an analysis of the historical and conceptual set of gendered dualisms within Western culture. These gendered dualisms begin from an idea of culture or society being separate and above nature and involve the identification of women and female with 'nature' and the 'natural'. The chapter then proceeds to discuss three of the main forms of ecofeminist social theory, essentialist or spiritual ecofeminism, materialist ecofeminism and political economy, and finally resistance ecofeminism.

Taking economics as a form of social theory, Chapter 6 looks at the ways in which economic theory has viewed, valued and conceptualised the environment. Beginning by showing how 'the economic problem' sets up the conceptual relationship between economics and the environment, it then outlines some of the historical connections between them in various economic schools of thought. In particular, the ways in which the emergence of the modern market economy and economics conceptualised and legitimated particular uses of the environment is discussed in terms of the relationship between land, labour and the enclosure movement, and the relationship which existed both historically and theoretically between material progress, poverty and economics. It then examines the relationship between economic theory and environmental policy-making within the contemporary liberal democratic state, before looking at environmental economics as a contemporary form of 'economising the environment'. Finally, it introduces and discusses ecological economics as a recent development within economics which attempts to integrate ecological and social insights into the examination of the economy.

Chapter 7 explores two recent social theories and how they analyse the environment – namely, 'risk society' and postmodernism. In the first half, Ulrich Beck's 'risk-society' thesis is discussed as a particular approach to the environment and environmental risk. The character of risk is explored before moving on to how the precautionary principle (which is fast becoming a central regulating principle of social-environmental interaction) can be seen as a logical extension of Beck's thesis. 'Reflexive modernisation' (a theme Beck shares with Giddens), stemming in part as a particular way modern societies can cope with increased environmental (and other) risks, is discussed in terms of how it seems to imply a redefinition of progress. A central part of this redefinition involves the extension of democratic forms of decision-making to more areas of social life. The latter half looks at postmodern approaches to the environment and environmental issues. Environmentalism and its commonalities with postmodernism are discussed in terms of a shared rejection of modernity. However, some problems with postmodern environmentalism are outlined. Primarily, an argument is made that environmentalism in particular and social-environmental problems in general, do not necessarily call for the rejection of modernity. Instead, and in keeping with one aspect of Habermas's thought, a claim is made that environmentalism can be seen as a critical analysis of and call for the fulfilment or completion of the 'project of modernity'.

Chapter 8 is an exploration of some of the issues involved in the relationship between ecology, biology and social theory. A critical analysis of sociobiology is offered as an example of the way in which biology and particular understandings of the natural world and human nature have been used to advance or support particular political positions. Using the work of Ted Benton and Peter Dickens in particular, the implications of ecology and biology for social theory are discussed in terms of the desirability and necessity for a 'critical naturalistic' form of social theory. Of particular interest in this and the concluding chapter are the consequences for social theory of seeing human beings as both 'biologically embodied' and 'ecologically embedded'.

Chapter 9 outlines some of the main principles of green social theory and uses this as the starting point for an examination of the 'greening' of social theory. Building on the insights of green social theory, and ideas discussed in previous chapters, some suggestions are made concerning the implications of the greening of social theory. A central claim in the greening of social theory is held to include the overcoming of a strict and permanent separation of 'society' and 'environment'. A consequence of this is that the greening of social theory requires an interdisciplinary approach to the study of society and environment in which the insights of the natural sciences are integrated with the insights of the social sciences. This in turn suggests new models of social theory and modes of social theorising.

# 1 'Nature', 'environment' and social theory

- What is social theory?
- Environment, nature and the nonhuman
- Social theorising and the environment
- The uses and abuses of the environment in social theory
- Four environments for humans in social theory
- The 'reading-off' hypothesis

## Introduction

What does one consider when one thinks about the 'environment'? Is the environment the trees, plants, animals that we see around us? Is it the Amazonian rainforests or the world's climate systems upon which all life on the planet depends? Are genetically modified organisms part of the environment? Is the environment the same as 'nature'? Does the 'environment' have to do with concepts such as 'biodiversity', 'ecosystems' and 'ecological harmony'? Can we say that the room where you are probably reading this book constitutes an 'environment'?

The problem (which can also be an advantage) with the concept 'environment', like many other concepts such as 'democracy', 'justice' or 'equality', is that it can take a number of different meanings, refer to a variety of things, entities and processes, and thus cover a range of issues and be used to justify particular positions and arguments. While of course the environment cannot refer to anything (that is, it refers to some identifiable and determinant set of 'things') it is an extremely elastic term in that there are many things – the room you are sitting in, the book itself, the chair, the desk, other people, the fly on the window, and the unseen micro-organisms and the air around you – all of which could be considered to constitute your present and immediate 'environment'. Like many things, the environment can mean different things depending on how you define and understand it, or who defines it.

In many respects thinking about and theorising the environment is one of the most enduring aspects of human thought. For example, the question of the proper place of human society within the **natural order** has occupied a central place in philosophy since its beginnings. *Hence, why, how and in what ways the environment, and related concepts such as 'nature' and the 'natural', are used in social theory is not only extremely interesting but absolutely crucial, given the different meanings and power of these terms when used in argument and justification.* For example, calling something something 'natural' implies that it is beyond change, immutable, fixed and given. Hence the power of using this term to justify a particular argument, and the need to be aware of how and why the environment and related concepts are employed in social theorising.

At the same time, alongside the theoretical or academic interest, there is a very important practical aspect to thinking about and theorising the environment in relation to society. This has to do with the increasing quantity and quality of environmental problems which face every society on earth, both nationally and globally. Global warming and climate change, deforestation, desertification, pollution, **biodiversity** loss and the controversies over the benefits and dangers of **genetic engineering and biotechnology** – all are familiar terms in our everyday lives. All of these, and others, seem to suggest that there is an 'environmental crisis' which faces humanity (and the nonhuman world), the like of which is unprecedented in human history. For the first time in history, humanity has at its disposal the capacity radically to alter the environment (primarily through the application of science and technology), and even has the capacity (though thankfully still not the willingness) to completely destroy life on earth 'as we know it' (as Dr Spock would say) through the collected nuclear, biological and chemical weapons of mass destruction possessed by a minority of the world's nations. At the same time as being the first generation which has this capacity to affect the environment, one could also say that (particularly since the rise of the green or environmental movement) this is the first generation which knows (or at least has some sense) that it is transforming the environment in a way which will affect the state of the environment inherited by future generations. Hence thinking about or analysing the role or place of the environment in social theory (the aim of this book) is not simply of theoretical, but also of great practical interest.

The importance of analysing the environment and social theory can also be seen when one considers that the majority of the world's environmental problems are largely the result of human social action or behaviour. Global warming, for example, is accepted by the vast majority of the world's scientists to be the result of increased carbon emissions by humans, principally through energy production and consumption (the burning of **fossil fuels**, such as coal, gas and petroleum to create electricity) and forms of transport which rely on such fossil fuels. Hence social theory, defined below as the systematic study of how society is and ought to be, has an important role to play in explaining, understanding and providing possible solutions to the 'environmental crisis'.

## What is social theory?

'Social theory' as a field of study is particularly difficult to accurately determine or define. As understood here, social theory is the systematic study of human society, including the processes of social change and transformation, involving the formulation of theoretical (and empirical) hypotheses, explanations, justifications and prescriptions. In disciplinary terms 'social theory' is often associated with sociological theory, and the origins of modern social theory have their origins in the sociological tradition. This book however, takes a broad rather than a narrow understanding of social theory, in that it encompasses sociological theory but goes beyond it to include other disciplines and intellectual traditions and approaches. As can be seen from the range of authors and disciplinary approaches surveyed in this book, social theory includes the 'social-scientific' approach to the study of society (in terms of the disciplines one finds in the social-scientific approach to studying society and social phenomenon – sociology and anthropology, politics, international relations, economics, legal studies, women's studies, cultural studies). However, social theory may also include the disciplinary approaches of history, philosophy and moral theory and cultural geography. Thus 'social theory' acts as an umbrella under which are gathered a range of approaches to thinking about society, explaining social phenomena, and offering justifications for advocating or resisting social transformation.

The main disciplinary approaches of this book are: sociological theory (including cultural theory), political theory, economics and political economy, but also includes the history of social thought. In broad terms what can be called an interdisciplinary conception of social theory is used throughout the book.

The historical origins of social theory can be found in the
**Enlightenment**, though aspects of modern social theory can be found in
pre-Enlightenment thinkers and schools of thought (particularly in
political philosophy and political economy, as outlined in Chapters 2 and
3). And it is in reaction to the Enlightenment and the emergence of
'modern society', that a large part of past and contemporary social theory
finds its subject. It is in the spirit of the early emergence of social theory
that a broad understanding of it is adopted here. In its origins, social
theory covered the broad field of the systematic or disciplined study of
society in all its various aspects, political, economic, cultural, social,
legal, philosophical, moral, religious and scientific. Social theory as the
systematic or scientific study of society included looking at such social
phenomena as the relationship between the individual and society, the
origins and character of cultural practices, and the relationships within
and between everyday life and social institutions, such as the family, the
nation, the state and the economy. As May points out, in the nineteenth
century the main trends in social theory were, 'First, an interest in the
nature or social development and social origins. Second the merging of
history and philosophy into a 'science of society'. Third, the attempt to
discover rational-empirical causes for social phenomena in place of
metaphysical ones' (1996: 13). In a similar fashion, this book attempts to
offer an equally broad and inclusive view of social theory, though of
course many issues, writers and ideas are necessarily left out, or only
briefly mentioned. At the same time, we can use the Enlightenment as a
way to demarcate modern social theory by noting that the 'subject' of
modern social theory is 'the analysis of modernity and its impact on the
world' (Giddens *et al.*, 1994: 1). In particular, modern social theory
analyses the impact of the industrial, liberal-capitalist socio-economic
system which has come to dominate the modern world.

Social theory typically has two dimensions, one descriptive the other
prescriptive. In its descriptive aspect, social theory *describes* society and
advances particular explanations for social phenomena, events, problems
and changes within society. For example, a social theory may involve
explaining the emergence of contemporary far-right politics across
Europe by reference to a rise in unemployment, the negative economic
effects of globalisation and a consequent appeal of populist nationalist
politics in response to the erosion of 'national sovereignty' or 'national
pride'.

The *prescriptive* dimensions of social theory are the ways in which social
theory not only tells a story of the way society is, but also tells how

society ought to be. Here social theory advances particular normative, or value-based arguments, justifications and principles to support its claims about how society ought to be ordered, changed or whatever. This prescriptive aspect of social theory can broadly take two forms. On the one hand, it can seek to justify the present social order, that is, suggest that the way society is is how it ought to be. This can be described as a 'mainstream' or 'conservative' position in which the aim of social theory is to legitimate, defend and justify the current way society is organised, its principles, institutions and ways of life.

On the other hand, some forms of social theory seek to argue that society ought to be transformed, organised along different principles and with different institutions from those upon which the current social order is based. These forms of social theory can be broadly described as 'critical' in that they are critical of the current way society is organised and seek to provide reasons for why it ought to be changed and organised along different principles or with different institutions. The classical example of a critical form of social theory is Marxism, which criticises the current capitalist, liberal democratic organisation of societies in the 'West' or 'developed' world, suggesting an alternative 'communist' or 'socialist' mode of social organisation. Below are some examples of mainstream and critical forms of social theory which will be looked at in this book.

| *Mainstream social theory* | *Critical social theory* |
|---|---|
| Conservatism | Marxism/socialism |
| Neoclassical economics | Feminism |
| Sociobiology | Ecologism/green social theory |
| Social Darwinism | Postmodernism |

While it is nearly always an advantage to adopt broad and flexible, rather than narrow and rigid, approaches to the study of society, such an approach is particularly advantageous (indeed, some would say necessary and not just desirable) when it comes to social theory and the environment. The adoption of an explicitly interdisciplinary approach to studying the relation between society and environment is something that has become a dominant perspective in recent work on this area (Barry, 1999), and will be discussed in greater detail in Chapter 9. In part, this is due to the rather simple fact that there is not just *one* relation between society and environment (as this and other books in the Routledge 'Environment and Society' series seek to demonstrate). Rather, the relation between society and environment denotes a series of relationships, physical, social, economic, political, moral, cultural, epistemological and philosophical, covering a multi-faceted, multi-layered, complex and dynamic interaction between society and environment. Given the multiple relations between society and environment, it is clear that no one discipline or approach can hope to capture the full complexity of the various relations between society and environment. Hence an interdisciplinary approach, drawing on a variety of sources is not only useful, but in some ways necessary in discussing the environment has been conceptualised, used and abused within social theory.

With regard to social theorising about nature or the environment (and as indicated later the two are not necessarily the same), one can trace two other approaches, alongside critical and mainstream. These are what one can call 'naturalist' and 'social constructionist' approaches. Naturalist social theorising about the environment generally takes the view that the environment is external to society and exists as an independent 'natural order' outside of society. Social-constructionist approaches on the other hand, see 'environment' and 'nature' as constructions of society, and therefore focus on analysing the internal relations within society. Combining them together, we get a fourfold schema of social theoretical approaches to the environment. This schema can be used as a rough way of understanding particular social theories and theorists.

| | |
|---|---|
| *Critical* | *Conservative/mainstream* |
| *Naturalist*  e.g. anarchism | e.g. Malthusianism/Sociobiology |
| *Social* | |
| *Constructionist*  e.g. Marxism | e.g. neoclassical economics |

**Figure 1.1** *Social Theory and the Environment Schema*
Source: Author

## Environment, Nature and the Nonhuman

One way of starting our exploration of the place or role of environment in social theory is to look at what we mean by 'environment'. Firstly, we can note that the environment is an 'essentially contested' term. The phrase 'essentially contested' simply means that the term has no universally agreed and singular meaning or definition. The importance of these issues should of course be obvious when social theorising about the environment and its relationship to human social concerns. One of the first and most obvious issues about the environment and social theory concerns the **fact/value** distinction. This refers to the way in which the environment, and related terms, are used not just in a descriptive sense, that is dealing with the facts, but how they are also used to express, justify or establish particular values or judgements, courses of action and reaction, policy prescriptions and ways of thinking. Thus while the environment is used to simply describe the world, that is tell us how the world is, it is also used to prescribe how the world *ought to be*, or making some normative (value) claim about something. For example, the term 'natural' carries with it a host of different value meanings, sometimes positive ones meaning 'wholesome' or 'healthy' (as in organic food), sometimes negative ones meaning 'uncultured' or 'backward' (as in passing judgement on a group's way of life).

A good way to start thinking about the environment is to list its various definitions and understandings. Often when one is trying to define terms or concepts, a good place to start is a dictionary and thesaurus. Here are some definitions of 'environment' that can be found:

environment:  'surroundings, milieu, atmosphere, condition, climate, circumstances, setting, ambience, scene, decor' (taken from a computer thesaurus).

environment:  'situation, position, locality, attitude, place, site, bearings, neighbourhood' (*Roget's Thesaurus*, 1988).

environment:  'surroundings, conditions of life or growth' (*Collins English Dictionary and Thesaurus*, 1992).

Thus while the environment is often taken to mean the nonhuman world, and sometimes used as equivalent to 'nature', it can take on a variety of meanings. The roots of the term environment lie in the French word *environ* which means 'to surround', 'to envelop', 'to enclose'. Another closely related French word is 'milieu' which is often taken to mean the same as environment. An important implication of this idea of environment is that 'An environment as milieu is not something a creature is merely *in*, but something it *has*' (Cooper, 1992: 169). What Cooper means by this is that environment is not just a passive background or context within which something lives or exists. It is also something that is possessed in the sense that to have an environment is an important part of what the creature or entity *is*. That is, to have an environment is a constitutive part of who or what the creature is, so that one cannot identify a creature without referring to its environment. On this reading, anything that surrounds or environs is an environment. But 'to surround' by itself tells us little. We need to know *what* is surrounded in order to know what *the environment* in question is. That is, without some specified thing to refer or relate to (a species such as humans, or a culture or place) the term 'environment' means very little. Or rather without a referent, the environment can mean everything that surrounds everything that exists. In referring to everything, it also refers to nothing in particular and is therefore of little use as an analytical concept for social theory!

Thus it is important to note that the environment is a *relational* concept or idea in that we need to know what or who is the subject of discussion in order to define an environment. While we may often speak of 'the environment', what is usually at issue is a 'particular environment'. Hence, often, but not always, the environment within social theory is defined in relation to ourselves and particular human social relations, and particular historical and cultural contexts. For example, when people in the Western world speak of the environment they usually mean the physical nonhuman environment, such as the countryside, forests, animals, rivers and so on. However, in other cultures the environment may include these things but also include non-physical things like spirits and the ghosts of one's ancestors. It is for this reason that it is misleading to equate the environment with 'nature' in the sense of the nonhuman natural world, though this is often how it is understood. For example, the 'environment' can refer to the non-natural environment, as in the human, social or built environment. At the same time, as a relational concept, we can speak of the environments of other animals, organisms or planets. It is an interesting and instructive thought to consider that we are part of the environment of other creatures (and of each other in many respects).

However, 'nature' does not only refer to the nonhuman world, but is, as Raymond Williams noted, 'perhaps the most complex word in the language' (Williams, 1988: 221). This is because 'nature' can and does refer to both 'human nature' and nonhuman nature (understood as natural environment), thus crossing the boundary between that which is human and nonhuman. Indeed, the complexity and power of 'nature' has to do in large part with the fact that it can be used to unite (as well as separate) the human and the nonhuman. Here are some meanings of nature:

Nature: 'Nature comes from *nature*, OF [Old French] and *natura*, L [Latin], from a root in the past participle of nasci, L [Latin] – to be born (from which also derive *nation, native, innate*, etc.)' (Williams, 1988: 219).

nature: '1.The essence of something . . . 2. Areas unaltered by human action, i.e. nature as a realm external to humanity and society. 3. The physical world in its entirety, perhaps including humans, i.e. nature as a universal realm of which humans, as a species, are a part' (Castree, forthcoming).

nature: 'n. inborn or essential character or quality; temperament, disposition; instinct; universe, especially of living things, collectively; unspoilt wild life, scenery, and vegetation; the original unaltered or uncivilised state, especially of man [sic] . . . [Latin. *natura*, from *natus*, past participle of *nasci*, to be born]' (*The Children's Dictionary*, 1969: 398).

These definitions point to the way in which 'nature' can refer to both human and nonhuman issues, properties, processes and entities. Thus we can say that every living thing (both human and nonhuman) has its particular 'nature', as in a more or less determinate set of innate dispositions, characteristics and impulses. At the same time, nature can also simply refer to the totality of the nonhuman world, making it synonymous with the natural environment.

However, sometimes nature opposes environment. For example, one of the enduring debates within social theory concerns the 'natural' or 'innate' causes of human behaviour as opposed to its 'environmental' or 'external' causes. This is the common 'nature versus nurture' debate one finds within social theory and everyday discourse. Here 'nature' refers to 'human nature' understood as some 'given' or unalterable *internal* essence of human beings, while 'environment' refers to the *external* social environment within which humans are brought up and socialised. This issue is dealt with in more detail in Chapter 8. Thus both environment and nature are extremely complex, contested as well as very flexible terms.

## Social theorising and the environment

In common usage, the environment usually refers to the physical world which environs or surrounds something. Most commonly of all, in modern parlance, the environment is often thought of as synonymous with the 'natural world' or 'nature'. That is, the environment is often thought of as something that is objective rather than subjective. This is another way of understanding the fact/value distinction in that to say the environment is objective means it is a factual reality independent of our subjective value judgements. As objective reality, the environment just is. Closing one's eyes or mind to one's surroundings does not mean that they disappear. This is something most of us learn as we grow older, young children often believe that simply closing one's eyes is sufficient to make their environment (and all it contains, such as angry adults!) go away. Now while I do not wish to suggest that the environment does not or cannot refer to 'nature' (meaning the nonhuman world and its processes and entities), a less restrictive understanding of the environment is a more fruitful approach to take when relating the environment to social theory. That is, thinking about the environment as something that can and does mean more than the 'natural world' can both help us in thinking about the natural world as well as revealing the complexity of social theorising about the environment.

One of the problems in social theorising about the environment has been that the latter has been viewed by the former as essentially something that is both nonhuman and also beyond human society and culture. So for example, the environment has been understood as the 'natural world' or nonhuman nature, something which surrounds us and is also beyond human culture. This is the view of the environment one gets from popular nature programmes on television, such as David Attenborough's 'Life on Earth' series. The point is not to reject these understandings but to widen how we think about the environment so as to incorporate these and other possible meanings. Using the term 'environment' as simply another way of speaking about 'nature' or the 'natural world' within social theory is understandable, but one needs to be aware of the danger of missing something important about the environment if we define (and thus confine) it so narrowly.

Particularly in modern everyday language and in modern social theory (Soper, 1995), there is a marked tendency simply to equate the environment with the 'natural'. Often one finds the two terms used interchangeably. An example is O'Brien and Cahn's statement that 'the study of nature, and the relationship between human civilization and the environment, have always held a prominent position in social and political inquiry. Humans have long been interested in discovering our place in the hierarchy of nature' (1996: 5). The point is not that we should never equate the two concepts – indeed it is very difficult to consistently distinguish 'nature' from 'environment' – but rather we should be aware that distinguishing between them is required in critically analysing the concept of environment within social theory. As in many forms of human inquiry (particularly in the humanities and social sciences) part of the process of theorising about something involves making distinctions between different concepts, terms, relations and processes.

One important distinction which can be drawn is between 'nature' as conveying an abstract, almost neutral sense of the nonhuman world, and 'environment' as associated with a more local or determinate sense of a nonhuman (or human) milieu or surrounding. That is, 'nature' is often understood as referring to the conditions of life (for both human and nonhuman species) and all that exists on this planet as a whole, while 'environment' is often associated with a particular subset of these conditions, a subset defined in relation to a particular organism or entity. So we can speak of 'nature' without referring to any particular organism or entity, but 'environment' implies the environment of some particular organism, species or set of these. As Ingold puts it, nature is the 'reality *of* the physical world of neutral objects apparent only to the detached, indifferent observer' while the environment is the 'reality *for* the world constituted in relation to the organism or person whose environment it is' (1992: 44). Or as Cooper expresses it, 'an environment [is] a field of significance' (1992: 170), that is, significant for someone or something. Even when both nature and environment are used in reference to the nonhuman world, 'nature' is often associated with an abstract, universal sense of the nonhuman world, referring to the totality of the latter. In contrast 'environment', refers to a particular, less abstract and more local and determinate part of the natural world.

Like many of our concepts and terms, environment and nature are formed in contrast to their opposites. As well as consulting a dictionary or thesaurus, another good way to get a sense of what terms mean is to seek out their opposites. At least at an initial stage of inquiry, one can find out quite a lot about a concept by seeing with what it is contrasted. This dualistic form of analytical inquiry simply means that we compare a particular term with its opposite. It is important to point out that this form of inquiry will not capture the full complexity of an issue, since thinking about something cannot be reduced to simply specifying something and then discussing its opposite, but it is a useful way to start. So to what do these terms 'non-environment' and 'non-natural' refer? What do they mean?

Since a whole book could be taken up in exploring the full range of issues involved in this task (see Soper, 1995), what I intend to do is highlight some of the more obvious ways in which our understanding of these terms can be advanced by comparing them with their antonyms. In the binary set of concepts below, we can find quite a lot out about the meaning of the environment (qua 'nature').

| | | |
|---|---|---|
| environment/nature | *opposite of* | human society/culture |
| nature/nonhuman | *opposite of* | human |
| naturally occurring | *opposite of* | human-made/artificial |
| nature | *opposite of* | nurture |

One of the first things we should note about 'nature' and 'environment' (when used as referring to the nonhuman world or processes) is that they are viewed in opposition to human society and culture. In this respect, whatever is environmental or natural is something which is separate from and independent of human society. And in some respects this seems to be, at least intuitively, true. For example, trees grow and ecosystems function independently of human society and culture. At this very basic level nature or the 'natural environment' does not depend on humanity. Indeed, the opposite would seem to be the case: that is, humans in common with every other living species, depend on their environment to survive and flourish. So on this first analysis: the environment is something that is separate from human society. *However, this separation does not mean that humans do not have a relation with their environment.* Since they depend on their environment, and exist within the environment, they obviously are related to their environment. But to say humans are related to and depend upon the environment is not to say that they are the same as the environment. Like any other species, humans exist is a condition of separation from but at the same time a relationship to and with their environment.

Secondly, there is also another dimension to the relationship: that between 'nature' and the 'natural environment' as 'nonhuman' in contrast to 'human'. 'Nature' as 'nonhuman' can thus be used to define what is 'human' or what is properly human. In this way, nature as nonhuman is an extremely important concept, one might say a foundational concept, in social theory, in that it defines what is the human, or properly human.

Thirdly, we can see that the 'environment' can refer to that which naturally occurs, in contrast to that which is human-made or artificial.

Indeed, this final set of opposing concepts – between the 'natural' and the 'artificial' – is one of the most central ways in which humans have and do think about the environment. We commonly think of the environment as entities (rocks, rivers), species (bears, lions, foxes) and processes (carbon cycles, hydrological cycles) which are emphatically not the products of human society. Thus the environment here is that which occurs without human intervention, and many natural processes and entities predate humanity and human society. Here we have a notion of the environment qua nature which is one of the oldest and enduring conceptions humans have of the environment. This particular conception of the environment resonates with the idea of the environment as something nonhuman, the external and eternal natural and naturally occurring surroundings which envelopes both humans and nonhuman entities.

In some ways, it finds an echo in the Christian doctrine of the environment as 'God's Creation', that is, the environment as something which is not of human origin or design. One can also appreciate the distinction between the 'natural' and the 'artificial' when one considers the difference people perceive between certain foods and goods which are 'natural' and those that are 'processed'. Added to a factual distinction between what is 'natural' and what is 'artificial' are a whole range of evaluative positions in which one or the other is seen as 'superior' or 'better' than the other. For example, Goodin (1992) suggests that a 'green' theory of value rests on the claim that naturally occurring processes have a particular value precisely because they are not the work of human hands. As he puts it, in answer to his question 'What is so especially valuable about something having come about through natural rather than through artificial human processes?' is that 'naturalness [is] a source of value' (Goodin, 1992: 30). That something is 'natural', of nonhuman origin, and existing independently of human actions or interests, is held by many people, to be something of value. According to Goodin, a 'natural' landscape is more valuable than a 'humanised' landscape, in the same way as a 'fake' or reproduction is never as valuable as the original. Placing such stress on 'natural' and naturalness is a distinctly 'green' position, though one which many non-greens may share.

However, on the other hand, there are those for whom the 'artificial' is superior to the 'natural'. Here we can think of arguments in which whatever is human-made or produced is viewed as more valuable than whatever is naturally created. An extreme example of this is what can be called the **technocentric** position which holds that human creations are vastly superior to natural ones, so that it can not only ask 'What's wrong with plastic trees?' but answer that there is nothing wrong with them, and indeed they are superior to natural ones. One can see some of the origins of this view of the superiority of the human over the natural in the 'perfectionist' justifications for human transformation of the natural environment that were prevalent in pre-Enlightenment Western Christianity, as discussed in the next chapter.

However, it also needs to be remembered that there is a continuum between the two poles of the 'natural' and the 'artificial'; there are of course many intermediate positions between them. As will be seen throughout this book, this distinction between the 'natural' and the 'artificial', and their relative evaluative weightings (superior/inferior) is something that shadows much social theorising about the environment and our relationship to it. One can trace many of the origins of the debates about the relationship between society and the environment through looking at how at different times and places, different values are attached to the 'natural' and the 'artificial'. For example, whereas nowadays there is a premium attached to things 'natural' (and not just for health reasons, as in organic food), not so long ago, natural produce was regarded as 'backward', 'uncivilised' or not advanced; a sign of socio-economic inferiority. For example, in the last century, to live 'close to nature' (either in hunter-gatherer communities or rural-agricultural settings) or consume natural produce, meant one was not as advanced or cultured as those who did not live close to nature (but in urban areas and cities) and who enjoyed 'artificial' and 'processed' goods and services. Thus there is no determinate or singular, agreed or fixed reading of the natural and the artificial; they mean different things and are given different evaluations in different social and cultural settings and in different historical periods. The point of social theory is to make us aware of these evaluative distinctions, to try and understand them, and if possible suggest explanations for them. In this way we can say that there are no 'value-neutral' readings of the environment as nonhuman nature. That is, when one describes the environment as nonhuman nature, implicit in those descriptions are certain value-judgements and normative positions. This is partly because 'nature' and the 'natural' carry with them various meanings and express a variety of evaluative judgements (ranging from the good/positive to the inferior/negative). And as will be seen later on, in discussing the 'reading-off' hypothesis, often when social theories 'read' the 'environment' they often project or map particular ways of thinking and values on to the environment rather than simply offering a 'neutral' or 'objective' account.

At a very basic level one can intuitively grasp what it means to say that the environment is **socially constructed** by noting how different societies, different ways of thinking and social theorising display distinct ways of thinking about and perceiving the environment. For example, while the environment for a typical city-dweller may mean the houses, buildings, waste spaces, parks as well as 'nature' (meaning the nonhuman natural world), for a country-dweller, the environment may mean fields, domestic and wild animals, hedgerows, stone walls and the seasons, as well as 'nature'. Thus environments differ and depend upon that to which one is relating the environment in question. At the same time, this example shows that while there may be different conceptions of the environment, they do not necessarily have to be contradictory. This in part is due to the relational character of the environment – that is, the environment is that which surrounds something, some entity or someone (including collections and groups). My environment (however I construct this) does not necessarily have to contradict your environment (however you construct it). The map (the representation), after all, is not the territory (the physical reality), but a particular 'reading' or representation of the territory. As Foster notes, '"The environment" . . . is something upon which very many frames of reference converge. But there is no frame of reference which is as it were "naturally given", and which does not have to be contended for in environmental debate' (1997: 10).

Alongside this discursive or conceptual sense of the 'social construction' of the environment, there is a *material* dimension to the 'construction' of the environment which refers to the real, material, physical production and transformation of the environment by the human species. Such human transformations of the environment by humans include agriculture, the creation of particular landscapes by human practices different from the environment if left in its 'natural' (i.e. untransformed) state, the creation of hybrid species of plants and animals as a result of human intentional selection and cross-breeding (which includes modern biotechnological techniques and the human manipulation of genetic information).

## The uses and abuses of the environment in social theory

Conceptions of the environment differ sometimes dramatically. In some cultures, or within particular worldviews (ways of thinking), the environment can include the dead, one's ancestors and/or other entities from the 'supernatural' realm, such as gods, goddesses, spirits, angels, ghosts, etc. Thus the environment, as that which environs, depends not only on something to environ, but what constitutes the surrounding environment. Hence the environment does not necessarily refer to the physical environment (whether natural or human-made).

The full complexity of the social construction of the environment can be seen if we examine how we think about the environment. 'The environment' as a term of social discourse (that is as a part of human language and thinking) is of course a human concept. It is difficult to imagine that other species see or construct their environment using the conceptual tools humans do. Indeed, the vast majority of other species do not 'conceptualise' their environment at all (at least as far as we know), they simply get on with it and live within their particular environment. To complicate matters, when we focus on human societies we find that there is often a difficulty in translating what one culture refers to as 'environment' into another cultural context or language.

One of the first things that we can say about the environment is that while it refers to something 'out there' in the 'real world', this does not mean that it cannot be a social construction. Even in scientific discourse (the theory and practice of the scientific community, with its internal principles, standards and rules for what constitutes 'scientific knowledge'), which is a discourse about 'facts' telling us 'how things are' in the environment, the latter can still be regarded as a social construction. The facts of science about the external world are human social facts, particular interpretations of the way the world seems to us. It is more or less accepted within the scientific community that there are no observer-independent facts about the world. The world we see around us, and the scientific knowledge we have of that world, is as it is because of the types of beings we are. That is, if humans could not see colours apart from black and white, other colours would not exist for us. Rather like Adam in the Garden of Eden (discussed later), humans 'name' things and in naming them they exist, or 'come into being' for us, which is not the same as saying we create them. For example, while entities such as trees would doubtless exist without humans, the concept or category of a 'tree' would not. In a world in which humans did not exist, 'trees' would not exist, though the vegetative physical entities to which the term refers would.

When we think about the environment and when we apprehend it, we do so from particular perspectives and in more or less distinct ways. One way of saying this is to say that humans (and nonhumans) have particular modes of apprehending the environment, that is distinct and different ways of seeing, feeling and thus 'constructing' the world (the world as it seems to them). For example, we can say that given the type of beings humans are, with our particular physiological and sensory make-up, the environment we see and appreciate will be different from, say, how a fly or a bat perceives the same 'environment'. However, this is not to say that all humans have the same environment. Clearly they do not, since humans live in a wide (though not unlimited) variety of different environments, ranging from the harsh arctic environment to the tropical environment of the Equator. At the same time, different cultural or other value-based views of the environment means that not all humans (who share the same senses) will necessarily have the same meaning of 'environment'. For example, conceptualising the environment as an 'inhospitable wilderness' carries with it different meanings and origins and will have different implications and effects from conceptualising it as a 'garden', or 'God's creation'. Some of the implications of this are bought out in the next section.

## Four environments for humans in social theory

### 1. Environment as wilderness

According to Rennie-Short, 'Wilderness is a word whose first use marks the transition from a hunter-gathering economy to an agricultural society' (1991: 5). That is, 'wilderness' is a view of the natural environment from a 'civilised' or 'cultured' perspective. Rennie-Short suggests that there are two general responses to wilderness. The first is a negative reaction, in which the dominant aim is to 'pacify', 'tame' or 'conquer' wild nature, turning it into a 'garden' for human enjoyment and in line with human aims.

This has been the dominant view of wilderness for most of human history, wild nature as dangerous, uncontrollable and unstable, a permanent threat to the human social order (Oelschlaeger, 1992). Thus in European history, wild areas, such as dark forests, mountains and swamps, were regarded as dangerous places, and people looked on them with fear. A good example of this is the status and view of forests in European folk-tales and fairy tales as places where humans were not welcome, which were inhabited by wild animals, such as wolves and bears, as well as 'supernatural' creatures such as witches, goblins, evil spirits and so on. Indeed, the root of the term 'bewilder', meaning confusion, disorder and overwhelmed, can be found in the idea of wilderness. People were in danger of becoming bewildered when they left the security of the town and entered the disorder and anarchy of the wilderness, which could then threaten the order of human society. This negative view of 'raw' nature is something which can be seen in the Christian idea of perfecting nature, taking raw nature and perfecting it, transforming it into a more productive, tamed and aesthetically pleasing environment. It is also something that can be seen in the history of America with its myth of the 'Wild West' and the frontier society it created, an historical experience which has left its mark on American national identity, as many authors have pointed out (Arnold, 1996; Nash, 1967). From this frontier society, sprang a society and mentality of 'rugged, free, independent individuals' pitting themselves against an untamed wilderness which they had to dominate and control. Thus how one conceives of the environment can be part of how one conceives of collective, especially national, identities.

The second, more positive view of wilderness, is what one may call the 'romantic' or 'green' view. Here wilderness is celebrated, something to be cherished and valued in the face of a world in which the natural environment is increasingly 'developed' or destroyed by humans. As Zimmerman puts it, 'Wilderness is a direct reminder that not every thing can be reduced to the status of a human product, project or construct; wilderness is the "other" which reminds humanity of its own dependency on the powers at work in wilderness, but also in humanity itself' (1992: 247). This view can trace its roots to the romantic reaction to the processes of industrialisation in the late eighteenth and early nineteenth centuries as discussed in Chapter 3. This positive view of wilderness can be seen in contemporary popular culture, in which 'nature wild and free' is not only viewed as something to be valued, but also under threat from the bureaucratic, mundane, and stultifying processes of **modernisation**. Popular films such as *Free Willy*, about the 'freeing' of a tamed killer whale to its rightful 'wild environment', express this contemporary sentiment.

## 2. Environment as countryside/garden

Environment as countryside represents a different sense of environment from wilderness. In opposition to 'wilderness' and wild nature, the natural environment as countryside can be seen as a 'garden', a 'tamed' or humanised natural environment. Here the environment denotes nature transformed by and in the service of human needs and aims. Historically and conceptually, one can situate environment as countryside between 'wilderness' and the urban environment. That is, the idea of the environment as countryside lies between the natural environment untouched by human hands (wilderness) and the created, artificial environment created by humans (the urban environment). The idea of the environment as countryside or garden is often used to explain differences in national culture and environmental attitudes between European countries and other 'industrialised' countries. Unlike other countries such as America and Australia, countries like Britain do not have areas of wilderness, and the natural environments of these countries are humanised environments in the sense that they are the product of past human transformation. They are more like worked gardens than wilderness.

Just as with wilderness, Rennie-Short (1991) suggests that there is a connection between environment as countryside/garden and national identity. As he puts it, 'In most countries the countryside has become the embodiment of the nation, idealized as the ideal middle landscape between the rough wilderness of nature and the smooth artificiality of the town, a combination of nature and culture which best represents the nation-state' (1991: 35). Echoing some themes which will be taken up in later chapters, he also points out that 'the countryside is seen as the last remnant of a golden age . . . the nostalgic past, providing a glimpse of a simpler, purer age . . . [a] refuge from modernity' (1991: 31, 34). For example, in Britain (as elsewhere) the countryside, its inhabitants and particular ways of life are often held to best preserve how life used to be, before crime, stress, competitiveness and materialism became daily features of life. Here the 'countryside' acts as a 'living reminder' of a 'better' age in which things such as patriotism, hard work, loyalty to the land and monarch, respect for authority, piety and community formed the substance of daily life.

At the same time, the idea and experiences of those who wish to escape the 'rat race' associated with life in the 'big city', 'drop out' of the 'mainstream' and 'return to the land' is something that has been a periodic feature of Western social theory and historical experience for the last 200 years. A good example of this is the 'counter-cultural' movement of the 1960s in Western societies, which as part of its alternative to modern life in industrialised, urban societies, advocated a rural, agricultural setting, living communally, frugally and in 'tune with nature'. The popular 1970s BBC series 'The Good Life' expressed some of these themes.

## 3. The urban environment

By the urban environment is meant the human-made spaces, buildings, developments and structures that one finds in towns and cities, as opposed to either wilderness (where there is little or no human trace) or the rural countryside (where the natural environment and its processes are 'managed' but not created by humans). The city and town represent the artificial environment that humans make for themselves and others, and the emergence of the urban environment, both historically and conceptually, is also the most modern of environments which humans (and nonhumans) inhabit. Indeed, the connection between the urban environment and modernisation is such that standard accounts of the latter have urbanisation as a constitutive aspect. That is, to be 'modern' is to live in an urban as opposed to a rural environment.

The creation and development of the urban environment over the last 200 years has profoundly affected how people viewed and thought about the natural environment. As more and more people moved from the land to the cities (as part of the 'agricultural revolution' which was a condition for the industrial revolution) they increasingly became removed from direct contact with nature. This removal from the natural environment in part led to a heightened sense of the symbolic status of the natural world (wilderness and countryside) and a concern for its preservation. That is, the less direct and daily contact people had with the natural world, the more symbolically powerful it became (as a symbol of 'a golden age' as in 'countryside' or a reminder of the 'disorder' which surrounded society, as in wilderness). As outlined in Chapter 4, in discussing Giddens, the experience of urbanisation and separation from the natural environment is a key aspect in the rise of environmental

**Figure 1.2 'Dudley Street: Seven Dials' by Gustav Dore (1872)**

*Source:* From London, Mansell Collection (in Clayre, A. (ed.) (1997) *Nature and Industrialization*, Oxford: Oxford University Press in Association with the Open University Press)

consciousness, the aesthetic appreciation of nature and concern with the preservation of the natural world. The positive value placed on all things 'natural' and 'environmental' we see today is in part due to the fact that most people do not have direct experience of nature, and the natural carries with it a sense of purity, goodness, wholeness and wholesomeness which many find lacking in the polluted, overcrowded, unsafe and artificial environments of the city and urban areas. In times of crisis or when people experience life as crowded, polluted, unsafe and so on, it is understandable why a return to a simpler, purer and less complicated way of life seems attractive.

## 4. The global environment

The idea of the 'global environment' is the most recent conceptualisation of 'environment', and has its origins in the debates about the 'environmental crisis' in the 1960s and 1970s. It is thus significant to note that the concept of the global environment originates against a background of vulnerability, threat and risk (some of which will be taken up in Chapter 7 in discussing Ulrich Beck's 'risk-society' thesis). While there has always been a global environment or biosphere, with the emergence of threats to it, it became an object of public and political interest. Of particular importance in the development of the global environment was the pictures of the earth taken from space in the late 1960s (see figure 1.3). It was these images more than anything else which really brought home to people that we all live on and share the same planet earth. From this developed an environmentally based 'planetary consciousness', expressed not just in the emerging green and environmental movements but also in events such as Earth Day in 1970, and the concept of 'spaceship earth' (Boulding, 1966).

However, it was in the 1980s and 1990s that the idea of the 'global environment' gathered momentum. With the emergence of 'global environmental problems' such as **biodiversity** loss, global warming, ozone depletion – that is environmental problems which are global in scope, though not necessarily global in origin – the idea of a global environment made perfect sense. The idea of the global environment has become a central aspect of social theorising about **globalisation**, with the global environment now taking its place alongside the 'global economy', 'global communications' and the 'global village'. Just as the global economy creates a global network of socio-economic relations between distant places and people, likewise the global environment expresses

**Figure 1.3** *Earth rising above the surface of the Moon, 1969*

*Source*: Courtesy NASA (in Pickering, K.T. and Owen, L.A. (1997) *An Introduction to Global Environmental Issues* (2nd ed.), London: Routledge)

another level of the interdependence of distant people and places. Environmental problems, like the modern global free market economy, do not respect national boundaries.

The global-environment idea carries with it some important messages. The first is interconnectedness, that is, we're all in the same boat when it comes to threats to the global environment. The second is that at its heart (and reflecting its origins) the global environment is an environment under threat from humans. The third, and following on from the last is that talk of threats to the '*global* environment' often carry with it the implicit idea that since these threats are global, faced by everyone on the planet, and we are all in the same boat together, then somehow we are all to blame. Yet, according to many greens and developing world activists, the latter is not the case, since the main cause of these global problems are in the advanced, industrialised world. However, the idea of a 'shared earth' which is related to the idea of the 'global environment' is used to convey the idea of shared fate and common responsibility of humanity.

## The 'reading-off' hypothesis

Perhaps the oldest and most dominant way in which the environment has been used in social theory is what can be called 'the reading-off hypothesis' (Barry, 1994). This hypothesis, as its name suggests, proposes that if we wish to find out about human society, there are important lessons to be learnt from a close examination of the natural world in general, and animal behaviour in particular. This hypothesis is based on the *power* of the notion of 'naturalness' in social theory and argument. An example of its power is how calling something 'unnatural' is a common way of ending a discussion or argument. To say something or someone is 'unnatural' is a pejorative, negative judgement, meaning they or their behaviour are objectively and simply wrong. If I do not like what you do, calling you or your behaviour 'unnatural' is a powerful put-down. Its power lies in that I do not have to argue or 'prove' how or why you or your behaviour is unnatural.

The reading-off hypothesis claims that we can both better describe or explain human society by applying the knowledge derived from the study of the nonhuman world, since we too are part of nature, a particular species of animal. It is thus common in naturalist forms of social theorising about the environment. But alongside this 'descriptive' claim, it also purports that the study of the nonhuman world can have prescriptive power. In short, the reading-off hypothesis within social theory states that we can describe the human social world as it is and prescribe how it ought to be from the application of knowledge gained from the study of the natural world. We can 'read off' how human society is and ought to be from looking at the nonhuman world.

The appeal of reading off from nature lies largely in the idea of the 'givenness' of nature and natural processes: that is, whatever is 'natural' or 'part of nature' is simply the way things are and ought to be, and there is nothing we can do to alter it. As Smith points out, 'The authority of "nature" as a source of social norms derives from its assumed externality to human interference, the givenness and unalterability of natural events and processes and behaviours' (1996: 41). Whatever is deemed 'natural' and 'unnatural' simply is, that is an unalterable 'fact' of the 'way things are' (and ought to be). Thus if one says that women *are* 'naturally' – i.e.

by their natures – mothers and home-makers, then one has a powerful argument for saying women *ought* to be mothers and home-makers and not seek employment outside the home. Or, if one says that homosexuality is 'unnatural', one is in effect saying that it is a transgression of or 'going against' nature and the natural order of things, and hence it is simply wrong and/or harmful. These uses of 'natural' and 'unnatural' are common in everyday and academic argument.

It is important to note that this 'reading-off' hypothesis is a two-way process. 'Reading off' how human society or social relations ought to be from an examination of the natural world inevitably involves the projection of social claims/aims/positions on to the natural world. Rather than 'read off' we 'read into'. That is, we 'read off' from the natural world what we project into it. Thus for some, the natural world is a place of harmony, cooperation and balance between different species and environments, while for others 'nature is red in tooth and claw', a place of competition and 'survival of the fittest'. In social theory, there are no determinate readings of the environment, because there are no value-free readings.

However, an important point that must be raised with regard to the reading-off hypothesis is not concerned with particular readings or interpretations of the natural world. Rather, we need to ask why this particular device or strategy is used within social theory in the first place, and ask why any reading of the natural environment should be seen as telling us something important about how human society is and ought to be. This theme of reading off from the natural environment will be explored in greater detail in the next chapter and Chapter 7.

## Conclusion

The environment, and its related terms, their meanings, status and significance for humans, and the relationship between them and humanity, constitutes one of the oldest themes in human thought. Particularly when environment is equated with 'nature' or 'natural' it is a powerful form of argument and thus one must be aware of how it is used (and who by) in social theory. In the next chapter we explore some of the ways in which early forms of social theory used the environment and related concepts in justifying or grounding particular positions and arguments.

## Summary points

- A broad, flexible and interdisciplinary understanding of social theory is used in this book. Social theory, as the systematic study of society, covers the following disciplines (with particular emphasis on the first three): sociology, politics, political economy, cultural theory, philosophy, cultural geography, legal studies and history.

- Such an interdisciplinary approach is argued to be particularly appropriate when dealing with the matrix of relations between society and environment.

- The environment and its connected terms, such as nature and the natural, are relational concepts.

- There are many meanings and understandings of the environment, but one of the most common is the equation of the environment with 'nature'.

- While this equation of the environment with nature can be useful, it is necessary to expand the concept of environment beyond a simplistic equation with nature.

- Within social theory, this idea of the environment as nature can be seen to express itself in the distinction between 'nonhuman' and the 'human', and the 'natural' and the 'artificial'.

- In thinking about the environment one needs to be aware of the social and cultural meanings attached to the environment. There are no 'value-neutral' readings of the environment.

- Some important conceptualisations of the environment include: environment as wilderness, environment as countryside/garden, the urban environment and the global environment.

- One of the most common uses of environment in social theory is the reading-off hypothesis in which how the human social world ought to be is derived or read off from how the natural world is organised.

## Further reading

On social theory, both Derek Layder's *Understanding Social Theory* (London: Sage, 1994) and Tim May's *Situating Social Theory* (Milton Keynes: Open University Press, 1996) offer good introductory accounts of the historical origins and main traditions or currents of Western social theory. A more advanced set of readings of contemporary social theory can be found in Anthony Giddens *et al.* (eds), *The Polity Reader in Social Theory* (Cambridge: Polity, 1994).

A difficult but worthwhile read on the various complexities of the idea of nature and the environment can be found in Kate Soper's *What Is Nature?* (Oxford: Blackwell, 1995). Also my own *Rethinking Green Politics: Nature, Virtue and Progress* (London: Sage, 1999) contains a fuller discussion of the distinction between 'nature' and 'environment' as well as the relationship between 'culture' 'society' and the former, and the 'reading-off' hypothesis. Also see David Cooper's 'The Idea of Environment', in David Cooper and Joy Palmer (eds), *The Environment in Question: Ethics and Global Issues* (London: Routledge, 1992) and Tim Ingold's 'Culture and the Perception of the Environment' in E. Croll and D. Parkin (eds), *Bush Base, Forest Farm: Culture, Environment and Development* (London: Routledge, 1992).

For an examination of the cultural dimensions of social-environmental relations see Kay Milton's readable and informative book, *Environmentalism and Cultural Theory* (London: Routledge, 1996). For an excellent and readable account of different ideas of the environment (such as wilderness, countryside and city) see John Rennie-Short *Imagined Country: Society, Culture and Environment* (London: Routledge, 1991) and Elizabeth Croll and David Parkin's edited volume, *Bush Base, Forest Farm: Culture, Environment and Development* (London: Routledge, 1992).

# The role of the environment historically within social theory

- Non-Western views of the environment
- The Judeo-Christian legacy
- The Enlightenment, environment and social theory
- The industrial revolution
- The democratic revolution

## Introduction

The aim of this chapter is to set the scene for the later discussions of the role of the environment in social theory by looking at how social theory has historically viewed and used the environment. This chapter traces some of the historical antecedents of how previous human generations at different times, places and within different cultures have conceptualised and thought about the environment and social-environmental relations. A second aim will be to look at some of the historical roots of Western social theorising about the environment in general, looking at the legacy of Judeo-Christianity and the Enlightenment in particular. Finally a third aim is to look at some of the historical origins of the 'green' social theoretical perspective, focusing on some antecedents of green thought in two broad reactions to the Enlightenment: namely, the reactions to the industrial revolution and the French and American 'democratic revolutions' of the late eighteenth century.

Historically, social theory has been largely concerned with reflecting on human society, critically analysing it, proposing the best arrangement of society for *human beings*. While there have been some notable exceptions, as will be discussed later, social theory has historically been overwhelmingly **anthropocentric**, that is, largely concerned with humans, human interests and human social relations.

One has only to examine some of the great texts of social theory (covering such disciplines as politics, philosophy, sociology, history, economics) to quickly see that 'the environment' as an explicit object of examination is either absent, or else seen as a natural 'backdrop' against which human history, politics and social development takes place. Thus the environment (largely viewed as 'nature') is often a mute or a passive object of human manipulation within the history of social thought. It is rarely at the forefront of social theory historically, being seen as something that just is, standing over and above human affairs and the enduring natural context within which those affairs occur. However, this is not to say that the history of Western social thought has little to say about the environment. For the most part the environment has been regarded as a necessary collection of resources or means to human ends, an attitude towards the natural environment which still predominates today, but which is being challenged by greens and others who suggest that this attitude is both morally objectionable and results in environmental problems for society.

## Non-Western views of the environment

While most of this book will concentrate on the relationship between Western social theory and the environment, where appropriate references will also be made to non-Western perspectives and insights. This is particularly important as there is a strong argument to suggest that many of the environmental and social problems that we see around the world today, at least in part, may have to do with the predominance of a particular Western way of thinking about and interacting with the environment. Here all I wish to do is indicate that there is and has been a variety of non-Western social theorising about the environment.

Historically, as in the Western world, most non-Western social theorising about the environment took religious and 'traditional' cultural forms (meaning that how people thought about the environment was largely governed by myths and stories which were handed down from one generation to the next as tradition). In the Middle Eastern civilisations of Egypt and Mesopotamia we can find a wide variety of ways of thinking about the environment. According to Hughes, 'The attitude of the peoples of Mesopotamia toward nature . . . is marked by a strong sense of battle. Nature herself was represented in Mesopotamian mythology as monstrous chaos, and it was only by the constant labours of people and the patron gods that chaos could be overcome and order established' (1994: 34). This idea of the 'struggle against nature' is something that has framed much of the debate about the relationship between human society and the environment, and is something that will be taken up in later chapters. It serves as an opposing view to the idea of human harmony with a bountiful nature which many suggest is typical of some aboriginal and 'hunter-gatherer' worldviews, which are discussed briefly below.

The Oriental religious teachings of Confucianism, Shintoism and Buddhism each had their particular views on the proper place of the environment in their particular worldview, and all had their own rules and principles concerning the treatment and use of the environment. Generally speaking, Buddhism displayed a marked respect for the natural environment. Hindu religious thought denoted particular ways of treating domestic animals, and forbade the eating of beef. Islam had its own particular set of rules, laid down in the Koran, about the proper way of thinking and relating the environment.

One of the things which all the 'great religions' of the world share (Buddhism, Islam, Judeo-Christianity and Hinduism), and which is important to note, is a common character as 'agricultural' religions. That is, these particular religions can trace their historical roots back to the period after (the majority of) humans had left the 'hunter-gatherer' stage of human social evolution and were prominent in societies and empires which were overwhelmingly agricultural civilisations. A second issue of note is that most of the civilisations in which these religions originated were also civilisations in which cities and towns were important places of political, economic, religious and military organisation and power.

Aboriginal peoples in Africa, Australia, the Americas and in Asia also had their own, usually spiritually informed, traditional ways of thinking about and treating their environments. The forms these traditional ways of thinking and acting took ranged from animism, a belief in spirits of the forest or of particular animals, nature worship and sun worship. In general, in marked contrast to the Judeo-Christian religious worldview (and also to the other formal religions outlined above), these traditional aboriginal cultures were less **anthropocentric** and more inclined to emphasise the continuity rather than the separation between the human and the nonhuman worlds. These more 'eco-friendly' worldviews have been a constant source of inspiration for green forms of social theory. As

Wall notes, 'Greens and fellow travellers have used existing hunter-gatherer groups and their ancient ancestors as an example of ecological good conduct' (1994: 20).

## The Judeo-Christian legacy

To the extent that theological debates about spiritual and worldly matters can be said to constitute a form of social theorising, we can say that Judeo-Christianity was a limited, though nonetheless significant, reflection on the relationship between human society and the natural environment. The importance of beginning our analysis of the relationship between Western social theory and the environment with an examination of the Judeo-Christian legacy cannot be overstated. For while many see Western societies and social theory as 'secular' or non-religious, it remains the case that exploring their Judeo-Christian origins and contexts can be extremely illuminating. Generally speaking, it is more accurate and useful to describe Western societies as 'post-Christian'. What is meant by this is that, although it is no longer the case that these societies are deeply Christian in the way they were in the past, it is still the case that Christianity and Christian values and perspectives have shaped and continue to be reflected in many of the practices, institutions and cultures of these 'secular' societies. This, as I hope to show below is particularly the case with regard to Western social theorising about the environment.

In the Jewish tradition, the 'natural environment' was generally seen as something akin to 'wilderness' and against which human society had to struggle. However, at the same time there are more harmonious views on the interaction of humanity and nature. It is worth noting that Judaism in particular, has much to say on the proper treatment of domesticated animals, and forbade the needless destruction of the environment even to subdue one's enemies (Swartz, 1996).

In contrast to some hunter-gatherer views of the environment, and more in keeping with the Mesopotamian one mentioned above, both the Jewish and Christian views of the environment were not of a 'giving environment' (Milton, 1996: 116–18). The idea of the 'giving environment' denotes the (often misleading) positive conception of the typical environment of hunter-gatherer peoples, in which people simply picked or procured what they needed from the abundant resources of their immediate environment, without much effort. In marked contrast the Judeo-Christian attitude to the environment is a combination of a negative view of 'wilderness' (viewed as chaos and a threat to human social order), coupled with a deep sense of how the environment required intensive human labour and effort, like agricultural and animal husbandry practices, in order that humans could survive and prosper from 'ungiving' and often hostile natural environments.

The latter idea of having to labour for a living in the world is directly related to the biblical story of Adam and Eve being banished from the Garden of Eden (which was similar to the 'giving environment' of some hunter-gatherer peoples), for having defied God and eaten fruit from the tree of knowledge. In banishing Adam and Eve from this comfortable environment in which all their needs were met without having to work (and in which they along with the beasts were vegetarian), God curses Adam and his descendants (i.e. all humans) to have to 'work by the sweat of his brow'. The importance of the Garden of Eden story, as the Christian creation story, is not whether it is 'true' or not. Rather its significance lies in its being one of the first systematic and most powerful stories or narratives about the relationship between humans and the environment. As such, we can say that this story constitutes an important attempt to theorise the environment and our proper relationship to it. It contains many of the elements that surface later in social theorising about the environment. These include: a particular conception of 'environment' and its status as the 'home' or 'proper place' of humans; the role of knowledge in how we think and ought to think about and interact with the environment; the distinction between a 'giving' and a 'non-giving' environment; the crucial role of human labour in our relationship to the environment; and finally, the dangers inherent in particular forms of thinking about and using the environment for humans. All of these, and others, are issues which arise in different forms of social theorising about the environment and will be discussed in later chapters.

In the Christian Bible one can trace some of the roots of how the environment has been viewed and treated within Western society and social theory. Typically, people point to the passage in Genesis in which God orders Adam and Eve to 'dominate and subdue' the earth and 'go forth and multiply' which demonstrates the extremely **anthropocentric** character of Christianity (see Box 2.1). This anthropocentrism within Christianity is an attitude to the nonhuman world in which the environment is viewed and valued instrumentally. That is, on this particular reading of the Bible, humans are permitted and indeed encouraged to use the environment and value it only insofar as it is useful to human ends or

## Box 2.1

## Judeo-Christian theory and the environment

'Then God said, "Let us make man in our image, after our likeness and let them have dominion over the fish of the sea, and over the birds of the air, and over the cattle, and over all the earth, and over every creeping thing that creeps upon the earth." So God created man in his own image, in the image of God he created him; male and female he created them. And God blessed them, and God said to them, "Be fruitful and multiply, fill the earth and subdue it; and have dominion over the fish of the sea and over the birds of the air and over every living thing that moves upon the earth." And God said, "Behold, I have given you every plant yielding seed upon the face of all the earth, and every tree with seed in its fruit; you shall have them for food"' (Genesis 1: 26-9).

### The domination of nature interpretation

'Christianity in opposing and destroying pagan animism made it possible to exploit nature in a mood of indifference to the feelings of natural objects' (White, 1967: 1206).

'[I]n the Christian separation of man from the animals and the Christian view that nature was made for man, there lie the seeds of an attitude to nature far more properly describable as "arrogant"' (Passmore, 1980: 12).

### The stewardship tradition

'Genesis . . . makes this duty [of 'man' towards all nature and all life] clear when it tells us that God put Adam into the Garden of Eden "to dress and to keep it", i.e. to manage and protect it' (Passmore: 1980: 29).

'The tradition of stewardship legitimates the reordering of the non-human world in the interests of human welfare provided this is balanced with a sufficient regard for obligations to conserve the natural world, to protect the moral interests of wild and domesticated animals, and to regard the interests of future generations as well as those of presently existing persons' (Northcott, 1996: 129).

'The end of man's creation was, that he should be the viceroy of the great God of heaven and earth in this inferior world; his steward . . . bailiff or farmer of this goodly farm of the lower world' (seventeenth-century Chief Justice, Sir Matthew Hale, quoted in in Passmore, 1980: 30).

---

### The perfection of nature thesis

'The word "nature" derives . . . from the Latin *nascere*, with such meanings as "to be born", "to come into being". Its etymology suggests, that is, the embryonic, the potential rather than the actual. We speak, in this spirit, of an area still in something like its original condition as "not yet developed". To "develop land", on this way of looking at man's relationship to nature, is to actualise its potentialities, to bring to light what it has in itself to become, and this means to perfect it' (Passmore, 1980: 32).

'The view that man has responsibility for handing over to his descendants a nature made more fruitful by his efforts is not . . . an entirely contemporary innovation, or an attempt to appeal to moral feelings which simply do not exist: it has deeper roots in Western civilisation' (Passmore, 1980: 32).

---

purposes. That is, the environment has no value in itself (**intrinsic value**), but has **instrumental value,** that is, its value or worth is given by how useful or instrumental it is in fulfilling some purpose other than its own, or the needs or ends of some other entity. According to White (1967), on his reading of Christianity, what he calls the 'domination of nature' story in Genesis makes Christianity the most anthropocentric of all religions. Indicative of the superiority of humans over the nonhuman world is not just that 'man' [sic] is created in the image of God, but God also gives Adam the power to name each creature.

However, in opposition to this 'domination' view, Passmore (1980) suggests an alternative interpretation of Christian teaching about the environment. According to him, there is a 'stewardship tradition' derived from Christian thought in which the natural environment was 'God's Creation'. This stewardship tradition predates Christianity according to Passmore, and origins for it can be found in the Greek philosopher Plato who in the *Phaedrus* wrote that '"It is everywhere the responsibility of the animate to look after the inanimate". Man . . . is sent to earth by God "to administer earthly things", to care for them in God's name' (quoted in Passmore, 1980: 28; see Box 2.1).

This was also the Jewish position. As Swartz points out, 'And though their efforts to tame the land, to make it more productive and more dependable, were often marvels of ingenuity, they understood, as well, the limits to their mastery – for they knew God as Sovereign of the Land, and . . . they acknowledged God's ownership' (1996: 88). Since the natural world had not been made by humans, it was not their exclusive property to treat and use as they wished. Within the stewardship tradition, rather than the nonhuman world being made for humans (a position which did eventually come to dominate Western views of the environment), as stewards of God's Creation humans were in a sense made for the non-human world, or rather they were God's 'managers' or stewards holding responsibility for God's property. This meant that there were certain rules governing how the environment, its plants, animals and so on were to be treated. One implication of the stewardship view and one which will be picked up in the conclusion, is that as stewards of Creation, humans had an obligation to pass on the natural environment to future generations (see Box 2.1). An important point here is whether this obligation meant passing on the environment in the same state as they found it, or whether humans were obliged to 'improve' the natural environment.

As Passmore (1980) and others (Pepper, 1984) have pointed out, this Christian stewardship view, in which social-environmental interaction was governed by religious considerations, began to give way to a more interventionist and anthropocentric viewpoint in which humans could use their God-given powers of creativity and ingenuity to 'perfect' nature for 'the Glory of God' (see Box 2.1). The 'perfection of nature' idea resulted in providing a religious justification for what we would now call the 'development' of the environment. This transformation of the natural world by humans in the West took many forms, from the creation of geometrically symmetrical landscape gardens of the famous eighteenth-century landscaper 'Capability' Brown (in contrast to the messy, irregular patterns of 'wild' or 'natural' environments), to the straightening of rivers and the draining of swamps.

Another extremely important contribution Christian thinking made to theorising the environment is the 'Great Chain of Being' (though strictly speaking this predates Christianity). The essence of this view, as the name suggests, was that the world was made up of a hierarchical set of relationships with God at the top of the chain and clay/dirt at the bottom, with angels, men, women, animals and plants in between (see Figure 2.1). Thomas Aquinas gave clear expression to this idea:

> As we observe . . . imperfect beings serve the needs of more noble beings;
> plants draw their nutrients from the earth, animals feed on plants, and
> these in turn serve man's use. We conclude, then, that lifeless beings exist
> for living beings, plants for animals, and the latter for man . . . The whole
> of material nature exists for man, inasmuch as he is a rational animal . . .
> We believe all corporeal things to have been made for man's sake.
>
> (Quoted in Kinsley, 1996: 110)

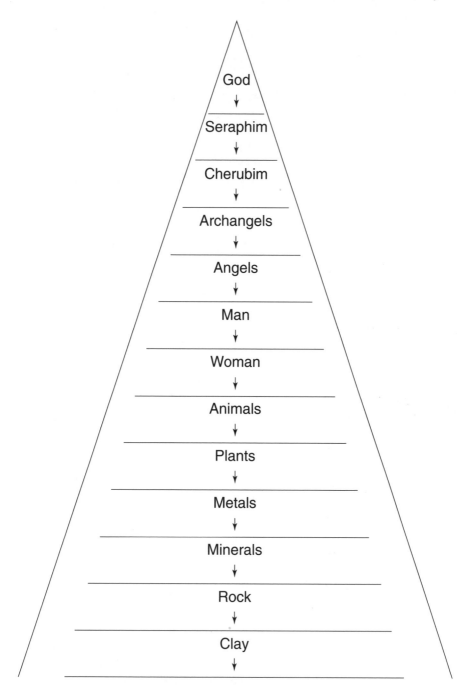

**Figure 2.1** *The Christian 'Great Chain of Being'*

*Source*: Adapted from Pepper, D. (1984) *The Roots of Modern Environmentalism*, London: Croom Helm

However, it is interesting to note that in both Jewish and Christian theology there have been those who have rejected this hierarchical view. For example, the Jewish kabbalist (mystic) Maimonides declared that, 'It should not be believed that all the beings exist for the sake of the existence of humanity. On the contrary, all the beings too have been intended for their own sakes, and not for the sake of something else' (in Swartz, 1996: 93). A similar argument was advanced by St Francis of Assisi who famously preached to the animals and developed a Christian pantheism in which the natural environment partook of the divinity and grace of God, and was not simply a set of spiritually meaningless or empty resources to be used.

Despite these and other counter currents, the dominant attitude of Judeo-Christianity to the environment has been one based on the Great Chain of Being. The hierarchical arrangement of entities implied different grades of value or importance such that those above were more valuable/important than those below. Often there are also gradations within these broad categories, such that certain metals, for example, were more valuable than others (gold was higher than copper), or within the animal category (cows were higher than rats). This divine order, in which there is a place for every living and non-living thing (both natural and supernatural), is something which still frames how many people view and think about the natural environment. Both in everyday life and in social theory we find that this Great Chain of Being idea operates, such that humans are regarded as 'higher' or 'more important' than animals or plants, and indeed this also led to views in which certain human beings were 'higher' than others (white European males being higher or superior to all others). As the pigs in George Orwell's *Animal Farm* put it, 'All animals are equal, but some animals are more equal than others'.

The Great Chain of Being idea linked with the 'perfection of nature' view in that those higher up the chain could legitimately transform and 'improve', 'perfect' or manage those entities below. Thus human transformation of the nonhuman environment was permissible within the framework of the Great Chain of Being. As the eighteenth century drew near the Christian legitimacy associated with this interventionist-instrumental view of the environment became increasingly difficult to sustain in the face of growing intellectual and practical challenges to the Christian worldview. Thus by the time of the Enlightenment in Europe and the beginnings of the industrial revolution in Britain, human use of the environment, particularly in agriculture, rudimentary commercial manufacturing, landscaping, forestry and waterways, had largely ceased to be legitimated by the idea of 'God's Creation' which implied that there were moral or normative limits to what humans could do to the environment. Human transformation and use of the natural environment increasingly became divorced from a strict medieval Christian framework from the sixteenth to eighteenth centuries. If the environment was becoming increasingly vulnerable to human manipulation and transformation from the sixteenth century on, this vulnerability became outright exploitation with the coming of the Enlightenment era in the eighteenth century. By exploitation is meant that the use of the environment was less and less regulated by moral (religiously based) considerations, and was viewed increasingly in non-moral, economic, terms, a consequence of which was an erosion of the boundary between legitimate 'use' and illegitimate 'abuse'.

## The Enlightenment, environment and social theory

'**The Enlightenment**' (which can also be termed 'modernity') is often understood as the series of interconnected and sometimes radical changes that took place within Europe in the mid to late eighteenth century, across numerous fields of human thought and action. There is no one exact date to which we can point to the dawning of the 'age of reason', nor is there one writer or school of social thought to which we can trace the exact origins of the momentous changes in European intellectual, political, economic and social life which occurred at this time. In terms of tracing and understanding the historical (and contemporary) relationship between social theory and the environment, the Enlightenment is of central significance. Not only do the origins of many present environmental problems lie in the Enlightenment (particularly the industrial revolution), but some of the roots of the 'green' critique and alternative to industrialism also lie in the various reactions to the Enlightenment. Hence the Enlightenment represents an important turning point in the place of the environment within social theory. As Porter points out, 'The Enlightenment believed people could improve themselves by improving nature, offering a programme of progress through science, technology and industry' (1994: 174). A typical example of the profound belief in progress and in the improvement of humanity by the application of reason (particularly scientific knowledge) in Enlightenment thinking is the following passage from Condorcet.

> A very small amount of ground will be able to produce a great quantity of
> supplies of greater utility or higher quality; more goods will be obtained
> for a smaller outlay; the manufacture of articles will be achieved with less
> wastage in raw materials and will make better use of them . . . The
> improvement of medical practice, which will become more efficacious
> with the progress of reason and social order, will mean the end of
> infectious and hereditary diseases and illnesses brought on by climate,
> food, or working conditions. It is reasonable to hope that all other
> diseases may likewise disappear as their distant causes are discovered.
> Would it be absurd, then, to suppose that this perfection of the humanity
> species might be capable of infinite progress?
>
> (Condorcet, 1995: 35–37)

The important point to note about the enlightenment is that human
progress and improvement is premised on the more effective exploitation
of the natural environment. As the passage from Condorcet above shows,
Enlightenment social theory had at its heart the exploitation of the natural
environment by the use of scientific knowledge and the application of
technology to industrial production. For purposes of exposition, what I
intend to do in this section is simplify the Enlightenment into two
component aspects, namely, the industrial revolution and the democratic
revolution.

## The industrial revolution

By the term industrial revolution is meant the various changes that took
place in European economic life both in terms of concepts, theories and
ideas and in terms of actual practice during the period from roughly the
sixteenth to the nineteenth centuries, which laid the basis for the
emergence and development of the modern industrial society. Britain is
often seen as the cradle of the industrial revolution, in that it was the first
country to transform itself along industrial lines. As the 'workshop of the
world', Britain exhibited many of the features which later industrial
societies would develop, and Britain became the model for
industrialisation.

Central to the industrial revolution is a particularly instrumental attitude
towards the natural environment. The environment was seen as a
collection of means for human ends, raw materials for the factories,
machines and new productive technologies which were being invented.
Science was seen as unlocking the secrets of nature, developing new

insights into its inner workings, and in conjunction with technology provided more effective ways in which humans could exploit it. In this way the natural environment became 'disenchanted', where once it was a meaningful order now under the cold, scientific light of reason, it simply became a collection of means (Barry, 1993). That is, whereas once the environment was variously seen as 'enchanted' (as in folk legends) or imbued with spiritual significance (as in Christianity where the natural environment was 'God's Creation'), with the industrial revolution, the environment was transformed and reduced to being a store of raw materials for human economic purposes. By 'disenchanting' the natural environment is meant the draining or eroding of meaning or significance from it, other than its status as a set of means for human ends. Other new forms of knowledge included the emergence of 'political economy' as the systematic study of the new capitalist economic system, with its novel free market, private-property-based economy. At the same time, the industrial revolution denoted radical changes in the type of economy and form of social organisation. The shift from a largely rural, agricultural economy to an urban, industrial economy created a new type of society, a 'modern' one, based on manufacturing, technological innovations and machinery in comparison to the feudal social order which preceded it.

Like all social change, the industrial revolution created 'winners' and 'losers' and was accompanied by social upheaval, unrest, suffering and pain. Chief among the 'losers' were the peasantry or 'commoners' who as a result of the enclosure movement (the privatisation of land to which 'commoners' once had rights of access) were forced into the emerging industrial urban areas and became the industrial working class. Another form of resistance to the industrial revolution were the 'Luddites' in the early nineteenth century (1811–16) who smashed machinery the introduction of which was causing unemployment and thus great social hardship. This will be discussed in Chapter 6.

As the industrial revolution continued there was also what one can call a Romantic and negative reaction, one that is particularly important for later social theorising about the environment in general, and 'green' social theory in particular. This Romantic backlash against the industrial revolution was motivated by how the latter was destroying and disfiguring the natural environment, turning once beautiful landscapes into ugly, overcrowded cities, polluting factories and mining operations. 'Was Jerusalem builded here, among these dark, satanic mills?' as the poet William Blake graphically put it in reference to the new mines, mills

Industrial capitalism launched the Modern Age, and ripped into the available raw materials with no regard for environmental consequences. By the 19th century Britain had become the 'workshop of the world'.

Having exhausted the supply of fresh trees the industrialists solved the fuel shortage with *fossilised* trees – coal.

The steam-engine churned. The cotton-mills hummed. The iron industry boomed. New coal-pits were opened. Towns grew into cities and village workshops into factories. Canals, roads and railway lines criss-crossed the country. Britain throbbed with industrial activity, to quote the school history books. But what else happened?

**Figure 2.2 *'Ecology and Industrial Capitalism'***

*Source*: Croall, S. and Rankin, W. (1981) *Ecology for Beginners*, New York: Pantheon Books

and factories which were disfiguring the English countryside, and which were at the heart of this new and polluting economic system. From the romantic perspective there was also something arrogant about human domination of the natural world. As Thomas Carlyle put it, 'For all earthly, and some unearthly purposes, we have machines and mechanical furtherances . . . We remove mountains, and make seas our smooth highway; nothing can resist us. We war with rude Nature; and, by our resistless engines, come off always victorious, and loaded with spoils' (in Clayre, 1977: 229). This theme of a 'war' against nature, is something that will also be discussed in Chapter 4.

## The democratic revolution

By the democratic revolution is meant the radical changes that took place politically during the late eighteenth and nineteenth centuries in theory and practice. The two key historical events here are the American Revolution (1775) and the French Revolution (1789). The main aspects of the democratic revolution concerned the principle of popular government – that is, government by the people, of the people and for the people – to replace rule by unelected monarchies, aristocrats and the church. The slogan of the French Revolution 'liberty, equality, fraternity' neatly sums up the essence of the democratic imperative of the Enlightenment, banishing the divine right of kings and the authority of organised religion. Other salient aspects of the democratic imperative of the Enlightenment included the use of the vocabulary of rights in political, social and moral thought, the increasing emphasis on the individual (both as citizen and producer/consumer); the emergence of representative government and liberal democracy; the establishment of constitutions, the separation of state powers and the rule of law rather than of man, and finally the creation of nation-states. Writers such as Thomas Paine, Jean-Jacques Rousseau, Voltaire, William Godwin, and Montesquieu wrote about, justified, supported and/or took part in this democratic revolution.

While less obvious than in the case of the aims of the industrial revolution, the democratic revolution was equally based on a particularly instrumental attitude towards and use of the natural environment. In the first case most supporters and theorists of democracy argued that this new form of government required material wealth (based on the exploitation of the natural world), and in this sense the industrial revolution was a necessary condition for the flowering of democratic politics. As Tocqueville noted, 'General prosperity is favourable to the stability of all governments, but more particularly of a democratic one, which depends upon the will of the majority, and especially upon the will of that portion of the community which is most exposed to want. When the people rule, they must be rendered happy or they will overturn the state: and misery stimulates them to those excesses to which ambition rouses kings' (1956: 129–30).

Secondly, the democratic revolution was a property-owning democracy in the sense that democratic rights were not extended to everyone. Only men with property were permitted to vote, and an important implication of this is that it served to further legitimate the idea of private property in land. That is, extending the right to vote implied extending private property in land to more and more people. This, of course, meant not only regarding the natural environment as raw materials but also as private and transferable property, which could be traded, bought and sold like any commodity in the emerging market economy. This property-based view of democracy was especially clear in the American case, largely because the American democratic revolution was strongly grounded in the political philosophy of John Locke for whom the goal of government was primarily to protect life, individual liberty and private property. Locke's ideas will be discussed in the next chapter. As the American Declaration of Independence puts it, 'We hold these Truths to be self-evident, that all Men are created equal, that they are endowed by their Creator with certain inalienable Right, that among these are Life, Liberty, and the Pursuit of Happiness'. According to Kramnick, in the context of the Enlightenment, 'Government's purpose was to serve self-interest, to enable individuals to enjoy peacefully their rights to life, liberty and property, not to serve the glory of God or dynasties' (1995: xvi).

Another dimension to this land-ownership issue was the special status of agricultural life and those who work the land. According to William Jefferson, one of the founding fathers of America, 'cultivators of the earth are the most virtuous and independent of citizens', adding, in a statement which echoes the anti-urbanism and anti-commercialism of some green thinking, that, 'Merchants have no country. The mere spot they stand on does not constitute so strong an attachment as that from which they draw their gains' (quoted in Miller, 1988: 207, 210–11). Echoing some of the issues raised in the previous chapter concerning the status of countryside and town/city, Jefferson thought that to dwell in the country was to dwell in virtue, while living in the city, separated from nature, was to risk corruption (Rennie-Short, 1991).

Thus the Enlightenment or modernity is an absolutely key moment in the relationship between social theory and the environment, since it represented a radical change both in theory and practice about how the natural environment was viewed, valued, used and conceptualised.

## Conclusion

This chapter has explored the historical relationship between social theory and the environment by outlining two opposing ways in which the environment has been theorised in Western social thought. On the one hand we have the religious approach of Judeo-Christianity, and the various ways in which it has theorised the environment and the proper relation between humans and the environment. These include the narrative of the Garden of Eden, the competing interpretations of the Christian attitude to the environment (namely the domination of nature view and the stewardship tradition), and finally the idea of the 'Great Chain of Being'. On the other, we have the profoundly secular approaches to theorising the environment offered by the Enlightenment, viewed as a combination of two revolutions: the industrial and democratic. An awareness of both Judeo-Christian and the Enlightenment are necessary as historical and conceptual legacies and frameworks to understand the character of social theory and the environment.

## Summary points

- Up until the Enlightenment, most social theorising about the environment took the form of religious, mythical or 'traditional' accounts of the origins of the natural world, humans, and the 'proper' relationship between the two.

- Analysing the Judeo-Christian worldview and its teachings are important for examining the historical origins of the relationship between Western social theory and the environment.

- There are competing views about the 'ecological' character of Christianity. On the one hand, there is the 'domination of nature' interpretation of Genesis in the Bible. On the other, there is the 'stewardship' view where humans are stewards or caretakers (not owners) of 'God's Creation', and also the 'perfection of nature' thesis, where humans are obliged or encouraged to 'perfect' or 'develop' nature.

- The Enlightenment (or 'modernity') marks a decisive change in how European civilisation thought about and used the natural environment. With the advent of the industrial revolution, nature became 'disenchanted'.

- The historical origins of social theorising about the environment can be traced to the industrial and democratic revolutions, and reactions to them.

## Further reading

For authoritative and scholarly accounts of the theoretical history of Western society, social theory and the environment see John Passmore's excellent *Man's Responsibility for Nature* (London: Duckworth, 2nd edn 1980) and Clarence Glacken's magisterial (and extremely long!) *Traces on the Rhodian Shore* (1967). A more focused account of the history of thinking about the environment can be found in Donald Worster's *Nature's Economy: A History of Ecological Ideas* (Cambridge: Cambridge University Press, 2nd edn 1994). Keith Thomas's *Man and the Natural World: Changing Attitudes in England 1500–1800* (Harmondsworth: Penguin, 1983), is a very readable account which looks at some of the historical changes in England that preceded the Enlightenment.

A useful overview of the historical place of the environment within Western thought is Derek Wall's reader, *Green History* (London, Routledge, 1995), which contains many edited original articles. David Pepper has written two books which cover some of the issues discussed in this chapter: *The Roots of Modern Environmentalism* (London: Croom Helm, 1984), and *Modern Environmentalism: An Introduction* (London: Routledge, 1996).

An excellent anthology exploring the relationship between religion and nature is Roger Gottlieb's edited volume, *This Sacred Earth: Religion, Nature, Environment* (London: Routledge, 1996), while for a more in-depth analysis of the Christian perspective see Michael Northcott's *The Environment and Christian Ethics* (Cambridge: Cambridge University Press, 1996), and Michael Barnes (ed.), *An Ecology of the Spirit* (Lanham, Maryland, University Press of America, 1994).

For an introduction to the some non-Western approaches to theorising the environment, see Peter Marshall's *Nature's Web: Rethinking Our Place on Earth* (London: Cassell, 1995), or J. Baird Callicott's *Earth's Insights: A Multicultural Survey of Ecological Ethics from the Mediterranean Basin to the Australian Outback* (Berkeley: University of California Press, 1994).

**The uses of 'nature' and the nonhuman world in social theory: pre-Enlightenment and Enlightenment accounts**

## Introduction

The aim of this chapter is to outline some of the ways in which the nonhuman world, its entities, processes and principles have been used (and abused) within the history of social theory. Of particular importance is the ways in which social theorists have appealed to some notion of the 'natural order' or 'nature' to justify, legitimate or illustrate their theories about and prescriptions for the social order. Just as religious thought both in Europe and elsewhere looked to nature for metaphors and lessons for illustrations of God's laws or plans for humans, modern social theory since the Enlightenment has also made both positive and negative references to the nonhuman world. The idea of nature as a 'text' (such as 'God's book', or like a book from which we can read, if we know the language of nature) or source of meaning (as opposed to simply a store of means to human ends) is something that has a very long history in human thought.

## Pre-nineteenth-century social readings of 'nature'

The 'scientific revolution' of the sixteenth to eighteenth centuries constitutes one of the significant backdrops against with social theorising about the nonhuman world took place. This period of European history is also concurrent with the beginnings of modern industrial capitalism, as Pepper (1996: 124) notes. The revolution in how the natural world was understood, its codification and use of mathematical methodology, rigorous modes of inquiry, and what one can call the scientific project of 'dominating' nature and finding out *her* secrets (in Francis Bacon's terms), constitutes the genesis both of 'modern science', and the modern 'worldview'. Unlike the pre-modern, medieval worldview which was based on Christian cosmology, the idea of a 'Great Chain of Being' discussed earlier, and an organistic view of the earth, the 'modern' worldview was that the earth is like a clock which is understandable and intelligible to human reason by the use of scientific modes of inquiry. This *mechanistic* conception of nature was at the heart of the emerging modern worldview, and laid the grounds for the Enlightenment and the industrial revolution. Indeed, this mechanistic view is still alive today, and can be seen in advertisements such as BUPA's (a British private medical insurance company) which states that 'Your body is a wonderful machine'. The important point to note is how attitudes towards the nonhuman world were intimately connected to particular ideas about progress, social and historical evolution and arguments about the best organisation of society.

## Forerunners and critics of modernity: environment, 'the state of nature' and social theory

Within political theory from the sixteenth century on, the idea of a 'state of nature' was a common device used to illustrate arguments about how the social order ought to be arranged, what the 'good society' is, and how political and social change should be viewed. This section looks at three important political thinkers to illustrate some of the different ways in which the environment was theorised in early social theory.

The 'state of nature' was a device used by various theorists to illustrate, explain and justify their particular social theories. Typically, the state of nature referred to a stage of human evolution was a 'pre-social' state, that is, how humans were prior to the creation of society, the state, social institutions, rules, principles and regulations. Generally speaking this stage of human evolution was regarded as inferior to and depicted a lower level preceding the 'civilised' stage at which human society is created and institutions such as the state, the monarchy (and market) are founded.

It implied a number of related propositions:

- there is a natural order, with a corresponding 'natural law';
- this natural order can be seen in the rest of nature;
- humans ought to follow this natural order;

## Thomas Hobbes

For the seventeenth-century political philosopher, Thomas Hobbes (1588–1679), later nineteenth- and twentieth-century anarchist views of humans as *naturally* harmonious, cooperative and not requiring a coercive state to impose social order, based on a reading of nature as cooperative, would have been heresy. For Hobbes, life in the 'state of nature' was 'solitary, poore, nasty, brutish and short', not the natural harmony and cooperation that anarchists thought existed within human society prior to the creation of the state. In this pre-state stage of social development, argued Hobbes, 'human society' as such did not exist, property was not secure, and individuals unfortunate enough to exist in such a state were continually in fear of their lives, not able to make plans for the future, and completely lacking security.

In this way, for Hobbes the idea of 'natural society' was not something he regarded as a positive stage in human evolution (as it was for Rousseau), but rather was a primitive and what later social theorists in the eighteenth century would call 'rude state of society' (as in rudimentary or basic). That is, humans living in a natural state were living in an uncivilised and backward stage of social development.

## John Locke

John Locke (1632–1704), another English social and political thinker, had a more benign view of human social life within a 'state of nature'. However, he did suggest that such a rudimentary state of society was one which, while not perhaps 'nasty, brutish and short', as Hobbes had so graphically suggested, was very basic, poor and in need of improvement and advancement. The way he suggested this be done had profound effects on how the nonhuman environment was viewed by later social theorists. Without going into the complexities of Locke's theory, one of his main arguments was that humans could claim parts of the nonhuman environment as their own, private and exclusive property. In this way, Locke is one of the first theorists to rationally justify an instrumental attitude towards and valuation of the nonhuman environment. In his view, the nonhuman environment, untouched by humans, is 'valueless'.

So coupled with Hobbes's view of the 'state of nature' as something to be viewed with horror, Locke's view of nature as worthless in its untouched state, together produced the dominant attitude of social and political theory to the nonhuman environment up until the modern era.

## Jean-Jacques Rousseau

However, in Jean-Jacques Rousseau (1712–78), we find a social theorist who differs from both Hobbes and Locke in his assessment of the state of nature; he is also one of the first and most powerful critics of the modern worldview, its social arrangements, aims, principles and animating goal of progress.

Going against the grain of Enlightenment thought, Rousseau argued that 'man' is naturally cooperative, and that 'primitive societies like those of the indigenous Americans and Africans were the "best for man"; civilization, far from being a boon, is always accompanied by costs that are greater than the benefits' (Masters, 1991: 456). Rousseau's idea of the 'noble savage', his positive view of life in the state of nature, made him an early critic of the Enlightenment and he can be seen as a precursor of later Romantic and green ideas of primitive peoples as paragons of ecological wisdom. By the 'noble savage' Rousseau, going against the received view of the status of 'savage' non-civilised peoples, suggested that this pre-civilised stage of human development, and human character, was in fact was more virtuous, morally good and admirable than the so-called 'advanced' civilised and cultured stage of social advancement. Writing in 1851, Rousseau declared that, 'Before art had new moulded our behaviours, and taught our passions to talk an affected language, our manners were indeed rustic, but sincere and natural' (1995: 365) and also that 'our minds have been corrupted in proportion as our arts and sciences have made advances towards their perfection' (1995: 367).

Rousseau was one of the first in this period to reverse the view that the 'Natural' was inferior to the 'Artificial', that a less complex society was necessarily inferior to an advanced civilised society such as those which existed in eighteenth-century Europe. Going against the 'progressive' spirit of the age, he viewed the emerging modern European societies in a rather different light from those social theorists, economists, philosophers and political theorists who viewed the Enlightenment as unqualifiedly a positive step forward in human social evolution. Rousseau questioned the 'progressive' character of civilised society as representing an advance over previous stages of human social evolution. In this he anticipated a key aspect of the Romantic critical reaction to the industrialism in the nineteenth century, and his critique of modernity and its conception of progress is something that is at the heart of the later emergence of green social and political thought in the twentieth century. At the same time, his use of an evolutionary framework anticipated the work of Charles Darwin in the nineteenth century, discussed later.

Rousseau's critique of the 'artificial' as the opposite of the 'natural', and his laying the blame for social ills on the corrupting effects of 'civilisation', constitutes one of the first critiques of the Enlightenment from a 'green' perspective. For Rousseau, nature and the natural environment represented innocence, authenticity and wholesomeness, against the corrupting effects of urban, sophisticated, civilised life. In this he was going against the prevailing notion of social evolution from primitive to agricultural, to city-states and on to the eighteenth century commercial society and modern nation-states, as expressed by the dominant forms of Enlightenment social theory.

In Rousseau's thought we can find echoes of the different values attached to different forms of environment in social theory, as outlined in Chapter 2. In depicting a rural, agricultural society of 'rough equality' and democratic government, one can see that he was using an idea of the environment as countryside/garden as morally (and politically) superior to the large, urbanised cities of his time. At the same time, his praise of the 'noble savage' implied a positive assessment of 'wilderness' and those who lived there.

## Nineteenth-century social theory and the nonhuman world

This section looks at the ways in which some of the main proponents of nineteenth-century social theory theorised about the nonhuman world.

But before proceeding it would be useful to briefly 'set the scene' by describing the intellectual and social context of the nineteenth century.

While there had been a romantic 'backlash' or reaction against the industrial revolution, as discussed in the introduction, this constituted a minority opinion. This Romantic reaction against the industrial revolution took the form of an aesthetic-cultural critique which highlighted the devastating effects of this revolution on the natural world and settled, traditional, rural ways of life. Leading proponents of this Romantic backlash included poets, writers and artists such as William Blake, Percy Shelley and William Wordsworth. Part of the rejection of the new industrial age rested on a suspicion and outright mistrust of the application of science and technology to the natural and social worlds. Like any great social change, the industrial revolution in Europe challenged old ways of thinking as well as of doing things, and equally like any change produced winners and losers, those that agreed and those who disagreed with the changes. By and large the nineteenth century could be called the first 'modern' century on account of the widespread changes in manufacturing, social life, culture, politics, economic organisation and above all else in the relationship between human society and the natural environment.

The application of scientific knowledge did not stop at the level of how humans could more productively exploit the natural world. In keeping with the Enlightenment belief in the ability of human reason to explain and solve almost everything, the nineteenth century also witnessed the rapid (and often indiscriminate) application of knowledge gained from the study of the physical world to the study of human society. The self-evident success and explanatory power of scientific knowledge convinced many social theorists that one had only to apply the methods of inquiry used in the natural science to the study of society. In this way the desire for the study of society to be 'scientific' led to the birth of the 'social sciences', and also led to the first attempts to 'read off' how society is, and ought to be, from observations of the natural world. By 'scientific' is meant the systematic study of phenomena, seeking universalisations, generalisable and law-like principles of explanation of cause and effect, using testable hypotheses which can be verified or falsified empirically by observation.

In the nineteenth century for a theory or theorist to be considered 'unscientific' was to be deemed as falling below the basic standards of social inquiry. When looking at the nineteenth century one must remember that most of the disciplines into which human knowledge and study are divided either emerged or took their present form at this time.

This was particularly the case with the 'social sciences'. Modern economic science, for example, emerged from the older tradition of classical political economy; sociology, 'the science of human beings and their behaviour' began to take shape towards the end of the century; while other disciplines such as law and history began to model themselves as 'scientific disciplines' and separated themselves from older disciplines such as philosophy and theology, which themselves began to develop into their own distinct academic disciplines. All in all the nineteenth century was a remarkable period at every level in Europe in ways of thinking about the natural and social worlds. In every area of social and individual life there were tremendous developments. In almost every aspect or dimension of human life you can think of – economic, social, political, cultural, religious, legal and personal – the nineteenth century in Europe was the birthplace of a new industrial society, which was both quantitatively and qualitatively different from any previous stage of human social development. And at the base of this industrial society, the ultimate source of its material success and self-understanding lay the domination of this society over the natural environment (and not just its local environment), principally as a result of the application of science and technology within new forms of socio-economic organisation combined with political and military power. Of particular importance was the emergence of the 'self-regulating' market as the principal means of organising the economy, which, together with the centrality of private property and production for profit, gives industrial society its 'capitalist' character (Polanyi, 1947; Goldblatt, 1996). This is discussed in more detail in Chapter 6.

While social theory was in the main tied to the aim of being 'scientific', what also unites different schools of European social thought and individual social theorists of the nineteenth century is a belief in the idea of social progress and the application of scientific knowledge as the means to achieve that end. With the 'death of God' as the philosopher Friedrich Nietzsche (1844–1900) announced towards the end of the century, humanity (or more specifically that portion of it resident in Europe and North America) and its progress became the ultimate aim or dominant goal of the new industrial society.

While Europe was experiencing the fruits of the industrial revolution, the rest of the world had its part to play in the unfolding drama. Though previous civilisations and societies had used 'non-local' environmental resources, often by force – that is, they were not self-sufficient in terms of the resources they required – the industrial society of the last century was the first human society to emerge that was truly 'global' in its reach and in its environmental impact. In other words, the industrial societies developing in Europe and North America, through the process of imperialism and colonisation of other parts of the world, Central and South America, Africa, India and South-East Asia, meant that this European industrial civilisation had at its disposal the environmental resources of almost the entire planet. From minerals and precious ores such as gold, silver and diamonds, to tropical hardwood timber, to spices, exotic animals, tea, coffee, bananas, and in the case of slavery, other human beings, Western industrial societies in the latter half of the nineteenth century required and depended upon more resources than their immediate European environment could provide. This is a point that will become important later on in discussing the green critique of industrial society. At the same time the 'discovery' of new lands, environments, climates, species and peoples had a direct effect on how the environment and environmental factors were articulated within social theory.

It is important to note the part played by ideas of environment and nature in justifying nineteenth-century colonisation and imperialism. For many European thinkers, indigenous peoples were seen as simply part of the environment, that is, they were not recognised as equal human beings. This inclusion of indigenous peoples as part of the environment is at the root of the *terra nullis* claim used to legitimate the colonisation of Australia. The logic of this argument is brutally concise. Since the Australian aboriginal peoples were part of the Australian environment, like other species, the land of Australia was uninhabited, 'terra nullis', a land empty of people and thus could be legitimately claimed as property.

Alongside these racist ideas were other equally objectionable views, such as the idea that peoples of 'hot environments' were lazy, backward and incapable of being industrious. Arnold, discussing the colonisation of India by Britain, notes how, 'Environmental forces – climate and disease above all – were repeatedly invoked to demonstrate and explain Indians' moral and physical weakness and to justify the indefinite continuance of imperial rule' (Arnold, 1996: 174). Unlike the temperate climates and environment of Europe, tropical environments produced work-shy, over-relaxed and non-energetic cultures and peoples. According to some social theorists of the time, 'nature' decreed despotism in Asia while the harshness of the Northern European climate made for hardy, self-starting industrious societies. Such environmental determinist ideas (that environment determines culture and psychological character) were quite common in the last century.

Describing the dynamics of this industrial mode of socio-economic life became the primary subject of analysis for social theory, which for the most part thought of itself as 'scientific'. However, as well as *explaining* the origins and principles of industrial society, how it functioned and developed, using scientific methods and modes of inquiry, social theory was also concerned with *prescribing* how this society ought to be. In keeping with the great explosion in human thought and knowledge that characterises this period, utopian plans, revolutionary critiques of and alternatives to the prevailing industrial social order, radical suggestions for social development and improvement, as well as less defences of the status quo, flourished within social theory. It is to an examination of how the prescriptive and descriptive claims of some of these social theories rested upon or used particular understandings of the nonhuman world, that we turn next.

## Progressive and reactionary Social theorising about nature

### Thomas Malthus

One of the most important areas of overlap between the environment and social theory in the nineteenth century begins with Thomas Malthus's theory of population. Malthus (1766–1834) criticised progressive Enlightenment thinkers such as William Godwin (1756–1836) and the Marquis de Condorcet (1743–94) at the end of the eighteenth century for thinking that the future of humanity was destined to be one of improved social, political and economic conditions. In the first edition of his infamous book, *An Essay on the Principles of Population as It Affects the Future Improvement of Society, with Remarks on the Speculations of Mr. Godwin, Mr. Condorcet and Other Writers*, published in 1798, Malthus takes the two named theorists and other Enlightenment thinkers to task for foolishly and unscientifically speculating that the 'era of reason', expressed politically in the success of the French Revolution in destroying government by unelected monarchy, gave any grounds for forecasting a society 'devoid of war, crime, government, disease, anguish, melancholy, and resentment, where every man unflinchingly sought the good of all' (Eklund and Herbert, 1975: 81). Anticipating the logic of the green 'limits to growth' argument of the 1970s (discussed in Chapter 9), Malthus argued that Enlightenment views on the future progress of society were unfounded. While he did think social progress such as less socio-economic inequality between classes was *undesirable* (something for which Marx was later vehemently to criticise him), Malthus's main claim was that the utopian visions of social progress suggested by Godwin (an anarchist) and Condorcet (a leading French philosopher of the Enlightenment) were *impossible* on 'natural' grounds.

Malthus 'had argued that the prospects for progress were continually threatened by population growth and the fact that food production could in no way match such growth' (Dickens, 1992: 22). The root of the problem for Malthus lay in the brute fact that while population increases geometrically (2,4,6,8), food supply increases arithmetically (1,2,3,4). The different rates of growth of the two 'proved' that social progress of the sort favoured and proposed by Enlightenment thinkers was impossible to achieve for what we can now call 'ecological' and 'biological' reasons. Because there were definite limits to the ability of the land to provide food without strict population control (and controlling what Malthus saw as the 'passion between the sexes' was something he thought was extremely difficult to achieve), social improvement would come to grief against these non-negotiable, nonhuman ecological limits. Because of this he argued that giving more resources to the poor will only increase their numbers (since he thought the poor were least likely to exercise sexual self-restraint), and thus only add to their misery. Hence he suggested that the state should not intervene to help the poor; rather, if state aid were removed they would be 'motivated' to find gainful employment (instead of being dependent upon the public purse) and encouraged to have fewer children.

Malthus's theory of population is not significant for its attention to 'ecological' considerations alone, but also for its attempt to be 'scientific'. While in the first edition of his *Essay* (1798), Malthus was largely offering some counter-speculations to those of Godwin and Condorcet, in subsequent editions (finally resulting in *A Summary View of the Principle of Population* in 1830), he attempted to give his theory some scientific credibility by using empirical evidence to prove the validity of his theory. This empirical evidence consisted of demographic, agricultural and other statistics and empirical data from different parts of the world, including travellers' diaries and other written accounts of journeys to foreign lands, and was flawed, patchy and would not pass statistical standards today. However, by the standards of early nineteenth-century social theory, his views were considered 'scientific'.

Malthus's social theory, in which he both described one aspect of modern industrial society and also prescribed particular policies, as well as basing it on 'scientific' principles of inquiry, can be said to constitute the model for social theory in the rest of the century. In particular, whether one was for or against industrial society, to be taken seriously one's theory had to be scientific, rigorous and if possible backed up by empirical evidence. For example, a clear indication of how many social sciences sought to base themselves on the natural sciences is August Comte (1798–1857), one of the founders of modern sociology, who stated in 1853 that, 'The subordination of social science to biology is so evident that nobody denies it in statement however it may be neglected in practice' (quoted in Dickens, 1992: 20). The implications of this for social theory are enormous: if the study of society could be reduced to biology, then we could both predict social behaviour as well as establish a set of scientific criteria by which we could judge what sort of social behaviour and social order we ought to have in accordance with scientific principles. The relationship between social theory and biology will be discussed in Chapter 8.

## Social theory and natural evolution: Darwin, Spencer and 'Social Darwinism'

Charles Darwin's (1809–82) theory of evolution by natural selection and Herbert Spencer's (1820–1903) extension of this principle to social evolution constituted the next significant juncture in the history of the relationship between the nonhuman environment and social theory. In particular there are two aspects of Darwin's theory of evolution which are worth noting. The first is his theory that humans were evolved from primates. This was an incredibly radical and controversial theory in its day, although, extreme Christian creationists aside, it is now the received scientific wisdom concerning human evolution. This theory not only went against the biblical story of the Creation of the Earth and humans by God, but, as will become clear later on, also served to undermine any strict separation between humans and the nonhuman world. Humans were not just *like* animals (somehow like them but actually 'higher' or 'superior' beings) and hence knowledge gained from the study of nonhuman animals might shed some light on human behaviour; after

Darwin one could say that humans *were* animals, a particular subspecies of primate, namely, *Homo sapiens*.

The second important aspect is Darwin's theory of natural selection which held that biological organisms were adapted to their environments. Due to a 'struggle for survival' between organisms it is those organisms which are best adapted to their environment which will survive and have more offspring and pass on their advantageous biological characteristics to their descendants.

Both of these aspects of Darwinian scientific theory were to have a great impact on social theory. In the history of social theory, it is Herbert Spencer who is seen as the main instigator of applying the 'Darwinian' principle of 'natural selection' and 'survival of the fittest' to the investigation of society. However, while Spencer is often credited with developing a theory of 'Social Darwinism', Dickens points out that, 'Spencer started developing his theories some time before the emergence of Darwin's *Origins of Species*' (1992: 20). On the other hand, Spencer took an extreme organic view of society (in that he saw society as a 'super-organism' and therefore subject to the same laws and patterns of development as any other organism), and used Darwin's discoveries regarding the evolution of species to explain, predict and prescribe social relations between human beings.

Social Darwinism as it developed from the mid nineteenth century on was a particularly harsh social theory in that, echoing the arguments of Malthus, it held that helping the poor only served to enable the 'unfit' to 'artificially' survive, and thus held back the course of social evolution which was premised on the survival of the fittest. In this way, Social Darwinism could be used to justify and legitimate a view of society in which there was little state interference in the 'natural' struggle for survival between human beings. It was, and sometimes still is, used to justify 'laissez-faire' capitalist or classical free-market forms of social order. Like Malthus before him, Spencer expressed a version of **classical liberalism** in rejecting state interference in society and economy, most explicitly articulated in his *The Man Versus the State* published in 1884. Spencer and Social Darwinism propounded what we would now call an extremely right-wing **libertarian social theory**. It proposed a particularly individualistic view of human freedom (the central principle or value for such social theories), based on economic competition and the free market. Social Darwinism also grounded its normative and prescriptive view of society and the proper relations of individual within it, on a particular organic conceptualisation of human society and of natural evolution by selection.

The study of natural selection and the interaction between and within species gave us a picture of 'nature, red in tooth and claw'. By analogy, social evolution and the interaction between individuals and groups in society, especially within the economic sphere, is and ought to be governed by the same laws of nature. Hence competition, self-interestedness, and ruthlessness in economic life could be justified as this simply conformed to the principles of social evolution and a particular conception of 'human nature' in which cooperation and solidarity were largely missing. It is perhaps no contingent fact that while Social Darwinism did flourish in Britain in the latter half of the nineteenth and early twentieth centuries, it was in North America, the land of 'rugged individualism', limited government and an extremely entrepreneurial culture, that Spencer's ideas and Social Darwinism took deepest root, particularly in the work of social theorists such as William Sumner (1840–1910) who stated that, 'The millionaires are a product of natural selection' (quoted in Dickens, 1992: 27). Like most Social Darwinists, Sumner's central political and social concern was with defending a particular, classical liberal view of the relationship between state, society and economy, in the name of a particular understanding of human liberty (Sumner, 1992). As Dickens notes, 'Early forms of social theory were largely constructed using analogies between societies and nature. Thus societies were seen as if they were developing like live organisms, or people were seen as struggling for survival in their environment, much in the same way as Darwin had specified in his theories' (1992: 56).

Many authors have argued that the application of Darwinian theory (or any knowledge of the natural world) to social affairs is not only misleading but, more importantly, serves an *ideological* function of legitimating particular forms of social arrangements, patterns of power and distribution of wealth. One such example is Social Darwinism. However, there have also been other social theorists who while following the same basic logic of Spencer (that is, reading off from the nonhuman world about how the human social world is and ought to be), have read a different story and drawn different lessons for human society from the natural world. At the other end of spectrum from Spencer's view of nature as 'red in tooth and claw' and his harsh message of the 'survival of the fittest' which underwrites the centrality of competition and struggle between humans within society (as well as between human society and the nonhuman environment) as necessary and desirable for social evolution and progress, we have the Russian aristocrat, social theorist and revolutionary anarchist Peter Kropotkin.

## Natural harmony: Kropotkin and the anarchist reading of the environment

In 1902 Prince Peter Kropotkin (1842–1921) published *Mutual Aid: A Factor of Evolution*, the main message of which was that cooperation was just as important as competition in both human and nonhuman evolution. For Kropotkin, and others who shared his political perspective, this of course led to a different set of principles and prescriptions for human society from those outlined by Social Darwinists, conservatives or liberals. For Kropotkin, as for most left-wing or communist anarchists, humans are not naturally selfish and engaged in a brutal struggle for survival; rather it was only under the particular prevailing social order that humans behaved in this manner. But more than that, Kropotkin's reading of natural selection, in contrast to Spencer's, was premised on the idea that humans were naturally cooperative and that if the 'artificial' institutions of the state and the capitalist organisation of the economy were abolished, humans could enjoy a more harmonious, cooperative and egalitarian social order. As Miller notes,

> Kropotkin . . . tried to show that Darwinian ideas, properly understood, could be invoked in aid of libertarian communism (thereby rebutting the normal implications of Social Darwinism). Beginning with animals and moving on to human societies, he argued that those groups which have proved most successful in evolutionary terms had done so by developing practices of mutual aid – practices whereby each member came to the help of others in need.
>
> (1991: 271)

In this way, Kropotkin and other communist anarchists echoed one of the main arguments that Rousseau had raised against the Enlightenment and the social and political arrangements of modern industrial society more than a century before. Rather than supporting a laissez-faire, competitive capitalist and unequal social order, social theory could find in nature a model of human society based on mutual aid, solidarity, equality and harmony. As Nisbet puts it,

> For Peter Kropotkin . . . the problem of community, that is, geunine and lasting community, resolved itself into a rediscovery of nature: not merely

the protection and proper development of nature in the external, physical and biological, sense, but also in the sense of seeking to build community – and in the long run society as a whole – on the most natural of interdependence among men.

(1982: 205–6)

This connection between external nature and internal human nature is something common to most social theorising about the environment which is in the form of 'reading off' social principles from nature. As de Geus, in his study of Kropotkin, puts it, 'in his graphic picture of an ecological society he takes into consideration a set of fundamental principles *which he believes can be derived from nature*, such as mutual aid, solidarity, cooperation, self-government, harmony, balance and community' (1999: 88; emphasis added).

## Marxist social theory and the environment

Karl Marx (1818–83) and Friedrich Engels (1820–95) and their political and social theory of Marxism is the next significant body of nineteenth-century social theory in which the environment played a particular (though ambiguous) role. In their famous work, *The Communist Manifesto* (first published in Germany in 1848), after stating that the bourgeoisie (the owners of capital, the ruling class in Marxist terms) has subjected 'the country to the rule of the towns' and 'rescued a considerable part of the population from the idiocy of rural life' (1967: 84), Marx and Engels go on to recognise the great achievements of industrial capitalism. Summing up the spirit of early social theory in the mid nineteenth century in terms of its acceptance of the domination of the nonhuman world as the basic premise of the industrial society, they state,

> The bourgeoisie, during its rule of scarce one hundred years, has created more massive and more colossal productive forces than have all preceding generations together. Subjection of Nature's forces to man, machinery, application of chemistry to industry and agriculture, steam-navigation, railways, electric telegraphs, clearing of whole continents for cultivation, canalization of rivers, whole populations conjured out of the ground – what earlier century had even a presentiment that such productive forces slumbered in the lap of social labour?
>
> (Marx and Engels, 1967: 85)

It is fair to say that historically classical Marxism, being a product of its time, did not address the range and significance of ecological issues that have come to play such an important part in late twentieth-century political and ethical discourse. Indeed, in so far as ecology stresses natural or absolute limits to economic development, early Marxist theory was vehemently anti-ecological. It is in the Marxist attack on Malthus's theory of population and his argument for subsistence wages that we can trace the predominant attitude of Marxist social theory to the nonhuman world (Barry, 1998c). For Marx, Malthus's theory was a piece of ideological justification masked as scientific inquiry. Malthus represented the interests of the landed classes and his theory of population was an attack on the urban poor and a defence of the political status quo.

Marx's attack on Malthus's ideas set the tone, and often the parameters within which the interaction between Marxism and concerns about the nonhuman world took place. For example, in this encounter are all the main ingredients which mark, and continue to mark, the relationship between ecological social thought and Marxism. First, there is the Marxist claim that stressing non-social ecological limits and conditions is anti-Enlightenment in general and anti-industrial in particular. Secondly, and following on from the latter, is the equation of anti-industrial with anti-working class, something which, for Marx, was abundantly clear in Malthus's work, and its main ideological aim. Thirdly, we have the importance of science and technology on both sides. On the one hand, we have Marx completely optimistic in the ability of technology, once free of capitalist relations, to transcend so-called 'natural limits'. On the other, we have Malthus's claim that his theory was fully supported by scientific and statistical data, which led to the opposite conclusion from that of Marx, and indeed was at odds with the dominant belief in progress that characterised the early development of industrialisation under capitalism, and which was reflected in social theory of this period.

In this way ecological concerns from a classical Marxist perspective was another fetter holding back the onward and inexorable rise of the revolutionary proletariat. In many ways ecological considerations were worse than bourgeois political economy because, unlike the latter, an excessive concern with the nonhuman world was held to be anti-industrial and anti-modern, and was argued to be driven by a desire to return to a pre-modern, agrarian, social order. When social theorists discussed the nonhuman world in terms of a Romantic defence of the natural world against industrialisation, against the 'disenchantment of nature' in Max Weber's (1864–1920) famous phrase, as in the Romantic poetry of William Wordsworth or Percy Shelley, or the social theories of Thomas Carlyle (1795–1881) or John Stuart Mill (1806–73), this merely confirmed the regressive and deeply conservative character of these social theories for Marxists. To be 'modern' and 'progressive' meant that nature had to be dominated, as Marx and Engels noted above, nature's forces had to be subjected to man. Those who objected to this domination and purely instrumental view of the nonhuman natural world were either simple-minded sentimentalists (poets such as Wordsworth) or reactionaries who were really motivated by a defence of a feudal, aristocratic social order based on a 'pre-industrial' and 'pre-modern' system of land-ownership (Malthus and Carlyle).

At the same time, it is not simply the case that Marx rejected the importance of the relationship between human society and the nonhuman world. Far from it. Marxist social theory is at root a materialist theory of human society, its dynamics and historical evolution. Unlike idealistic social theorists such as the German philosopher G.W.F. Hegel (1770–1831), the basic tenet of Marxism is that it is the material conditions and relations within a society which determines its character. It is not ideas that cause societies to change but rather the material relations under which the society organises its economic life which rather determines (or influences) ideas in society. The starting point for Marx is the brute fact that humans have to produce their own means of subsistence. What he means by this is that humans have to use their labour power, skills and creativity to transform the nonhuman world into the things, goods and services they need to survive. This, according to Marx and Engels, is what distinguishes humans from the nonhuman world. As they put it, 'Men can be distinguished from animals by consciousness, by religion or anything else you like. They themselves begin to distinguish themselves from animals as soon as they begin to produce their means of subsistence . . . The nature of individuals thus depends on the material conditions determining their production' (quoted in Parsons, 1977: 137). Though, like all other species, humans are dependent upon their environment for resources in order to survive, Marx held that humans were different from the rest of nature (and here he was simply stating the prevailing dominant view of the matter) because they did not simply take from nature whatever their natural environment afforded. Except in the 'primitive' hunter-gatherer stage of human evolution, the story of humanity was one where by their collective actions they transformed their environment, and by their labour power transformed the 'raw materials' of the nonhuman environment into useable and valuable artefacts: such as dwellings from simple huts to large cities, clothing from animal furs to designed fashions, and a whole range of goods, things and commodities (see Figure 3.1).

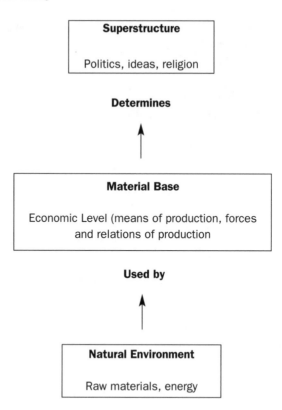

**Figure 3.1** *The Marxist Model of Socio-Historical Change*
Source: Author

Marx and Engels held Darwin in high esteem, and early Marxist thought agreed with the idea of evolution in history (as in nature) as being based on 'competition'. As Marx wrote, 'Darwin's book is very important and serves me as a basis in natural science for the class struggle in history' (quoted in Dickens, 1992: 45). That is, the Marxist model in which historical change, the evolution of society from one historical stage to the next, which was based on the idea of 'class struggle being the motor force of history', could find support in Darwin's ideas of natural evolution and competition between species.

Following Locke in many respects (though of course drawing completely different political and social lessons), Marxist social theory was premised on the idea that the nonhuman world if left to itself, unused and untouched by human hands, was 'valueless'. Whatever was of value in the world, was the product of human labour and creativity. This was the essence of the Marxist labour theory of value. The problem with the capitalist organisation of industrial society for Marxists was that the social organisation of this society was such that the vast majority of the people were denied the full fruits of their labour.

Because under capitalism, those who owned the capital, the factories, the machinery, etc., were in a stronger economic position that those who had simply their labour power to sell, the latter were, according to Marx, exploited. Thus, while the industrial **mode of production** (including the factory system, the extensive use of science and technology, and a complex division of labour), was premised on the intensive exploitation of the nonhuman world, its capitalist character also meant that the workers or the proletariat in Marxist terminology were also exploited. And while there have been attempts to claim that there are aspects of Marx's work in which there is a concern with overcoming the exploitation of the nonhuman world (Parsons, 1977), by and large we can say that the primary concern of Marxism was liberating human society from capitalist exploitation, inequality and oppression. In other words – and this is an important point for understanding much of the history and present state of the relationship between Enlightenment social theory and the nonhuman environment – *Marxism expresses the thoroughly 'modern' view that human social progress (and in this case 'liberation' from a capitalist exploitative social order), is dependent upon the exploitation and domination of the nonhuman environment.* As will be seen later in discussing critical theory (Chapter 4) and green social thought (Chapter 9), this is a position which has come under sustained attack in the latter half of this century.

The basic argument of Marxism as a representative modern social theory of industrial society was not about the great strides capitalism had made in finding the 'secrets of nature', and new ways of exploiting 'natural resources' which produced great amounts of material wealth, food, shelter and goods, as well as decreasing illness and increasing the average human life-span, and creating the most affluent social order the world had ever seen. Marx's problems with capitalism were not that he objected to the wealth-producing process, which was based on the exploitation of the nonhuman world; rather he argued that the fruits of this remarkable social order were not distributed equally because a few (the bourgeoisie or the owners of capital) enjoyed the gains while the many (the proletariat or workers) had to bear the costs and reaped few rewards. Marxism argued that the fruits of social labour should be shared equally amongst everyone, and not be the preserve of a privileged few who happened to own the means of production.

If nature was exploited under capitalism, Marx's views of a 'post-capitalist' or communist organisation of society was based on the hyper-exploitation of nature. His basic critique of capitalist industrial society was that its relations or production (particularly those relating to private property) were holding back what he called the forces of production (technology, science, the division of labour) from producing even greater levels of material wealth and affluence (see Figure 3.2). In other words, capitalism was not exploiting the nonhuman world efficiently enough to the fullest extent possible; nor did it distribute the wealth created in a just manner.

For Marxists, it is one of the great contradictions of capitalism that the social order which has put a man on the moon, is unable to eradicate poverty, homelessness and many other socio-economic problems. Marx's vision of a post-capitalist society is premised on the existence of 'material abundance': that is, communist society is one which has transcended material scarcity. Unlike all previous human societies in which there has never been enough material goods for everyone, and hence every social order has to have some principles or institutions for distributing what goods are available, Marx's communist society was one in which the principle of distribution was 'to each according to their needs'. In other words, Marx was envisaging that under communism the exploitation of nature and the production of wealth, goods and services would be so efficient and productive that all human material needs would be met. Whereas under capitalism only a few can afford an expensive car, for example, under communism everyone could have a car if they needed one because the economic exploitation of nature would be so efficient. Free of constraining, inefficient as well as illegitimate capitalist property relations (and the state, which would 'wither away'), human beings would in a position to use the world around them as they saw fit and to meet their every material need. Nobody would go hungry or homeless in this vision of a future post-capitalist society, and more than this, getting rid of capitalist relations of production would enable a more rational, planned, intensive and ultimately more productive exploitation of the nonhuman environment. Human liberation and emancipation were to be achieved at the price of the greater exploitation and instrumental use of nature.

A more 'environmentally sensitive' interpretation of Marxism can be found in some of the works of Engels, though the emphasis is on the urban environment rather than the natural one. In his *The Condition of the Working Class in England*, Engels described the degrading, filthy and

# The Marxist Boomerang

WAS MARX A GROWTH FREAK?

Marx thought you could fight capitalism by allowing it to grow, thus creating a strong class of industrial workers who would eventually 'break their chains' by kicking out the owners and taking over production themselves. He called such growth the development of the *productive forces*.

MAINLY TECHNOLOGY AND LABOUR

Most modern technology, however, by plundering nature is in conflict with the *conditions of production*.

NATURAL RESOURCES LIKE RAW MATERIALS, CLIMATE AND BIOLOGICAL FERTILITY

Marx left the impression that natural resources were in principle unlimited – that humanity would always find new ones. In the Communist Manifesto he declared that one of the aims of socialism would be 'to increase the total of productive forces as rapidly as possible'.

NOWADAYS THAT'S A RECIPE FOR ECO-DISASTER

**Figure 3.2** *'The Marxist Boomerang'*

*Source*: Croall, S. and Rankin, W. (1981) *Ecology for Beginners*, New York: Pantheon Books

unhealthy urban and working environments of the emerging urban working class, and suggested that the coming communist society would create less unhealthy, unsafe and more aesthetically pleasing urban, living and working environments. At the same time some recent theorists, such as Benton (1989, 1993) and Pepper (1993), working within the broadly Marxist tradition, have suggested more nuanced and less unecological forms of Marxist social theory (for an overview of some of these recent developments between Marxism and ecology see Barry 1998c).

As Giddens has commented, 'in general it is the case that Marx was not a critic of industrialism. Rather for him industrialism holds out the promise of a life of abundance, through turning the forces of nature to human purposes. It is a particular mode of organizing industrial production – capitalism – that needs to be combated, not the industrial order itself" (quoted in Cassell, 1993: 329). Hence Marx can be seen as wanting to intensify the exploitation of the natural environment which capitalism had begun, but end the exploitation of humans by humans and distribute the fruits of the exploitation of the environment more equally than under capitalism.

## Liberalism, Utilitarianism and J.S. Mill: the first 'green' social theorist?

The place of the environment in the work of John Stuart Mill (1806–73), one of the greatest liberal political thinkers of the last century, can be traced to his recovery from a mental breakdown while he was aided by reading the nature-loving Romantic poetry of Wordsworth and Coleridge. His highly original views about what we would call today 'ecological issues' were atypical of his time and anticipated many issues that would later come to be termed 'green' or 'ecological'.

For example, in an essay entitled 'Nature', Mill demonstrated an awareness of the foundational importance of nature and related ideas in social theory. Particularly interesting for our purposes, Mill in this essay also highlighting some of the dangers of the 'reading-off' hypothesis, outlined in the last chapter. As he puts it,

> the doctrine that man ought to follow nature, or in other words, ought to make the spontaneous course of things the model of his voluntary actions, is equally irrational and immoral. Irrational, because all human action

whatever, consists in altering, and all useful action in improving, the spontaneous course of nature. Immoral, because the course of natural phenomena being replete with everything which when committed by human beings is most worthy of abhorrence, any one who endeavoured in his own actions to imitate the natural *course of things would be universally seen and acknowledged to be the wickedest of men.*

(Mill, 1977: 311; emphasis added)

Thus for Mill, 'reading off' or following nature in how we ought to live and organise society is dangerous, irrational and likely to lead to immoral consequences.

While this essay illustrates that Mill did not have a benign view of nature (like Kropotkin) and tended towards seeing it as close to the image of nature being 'red in tooth and claw', this does not mean that Mill is dismissive of the importance of the natural world for humanity and social theory. As Stephens points out, 'Mill's vision of the natural world was not benevolent . . . but it was nonetheless mitigated by appreciation of its aesthetic and regenerative qualities, as his devotion to the poetry of Wordsworth and Coleridge testifies' (1996: 378).

The appreciation of the regenerative and aesthetic qualities of the natural environment, which has its roots in Romanticism, was a view shared and extended by American transcendentalism of the late nine-teenth century. This American movement in social and literary theory and practice, was a form of 'nature religion/spirituality' in that it saw God and spirituality as immanent in nature (especially the American wilderness). According to the transcendentalist view, the direct experience and appreciation of nature was a way to enter a 'higher' or 'transcendental' realm of eternal truth, beauty and happiness, away from the mundane distractions of our everyday, urban world. Thinkers associated with this movement, which was directly connected with the debates about the relationship between American national identity and wilderness, discussed in the last chapter, included Henry Thoreau and Ralph Waldo Emerson.

In 'Of the Stationary State', book IV, chapter VI of his *Principles of Political Economy*, first published in 1848, Mill suggested a radical critique of received wisdom concerning the progress of society. He suggested that the desire for more and more material goods and services, based on the domination of nature and the more intensive use and application of science and technology, was perhaps a too narrow view of 'social progress'. What Mill had to say is worth quoting in length.

> I cannot, therefore, regard the stationary state of capital and wealth with
> the unaffected aversion to it so generally manifested towards it by
> political economists of the old school. I am inclined to believe that it
> would be, on the whole, a very considerable improvement on our present
> condition . . . It may be a necessary stage in the progress of civilization
> . . . But the best state for human nature is that in which, while no one is
> poor, no one desires to be richer, nor has any reason to fear being thrust
> back, by the efforts of others to push themselves forward . . . It is only in
> the backward countries of the world that increased production is still an
> important object: in those most advanced, what is economically needed is
> better distribution, of which one indispensable means is a stricter restraint
> on population.
>
> (Mill, 1900: 453)

As Eklund and Herbert note, 'alone among the classical economists, Mill
did not believe the stationary state was undesirable, since . . . it provided
the necessary condition for his program of social reform' (1975: 120). In
articulating both a critique of the dominant view of social progress (as
the growth in material goods and services and undifferentiated economic
growth), based on a notion of natural limits, and in proposing an
alternative view of human social development which stressed the non-
material or qualitative dimensions of development, Mill in proposing the
stationary state economy anticipated some of the key concerns of green
social thought of the 1960s and 1970s. These ideas will be discussed in
the next chapter.

In many respects, taking together Mill's social theory, his view of
progress and the need for social reform via distributive policies, coupled
with what Macpherson (1973) has termed his 'developmental' strand of
liberalism (to be distinguished from the 'classical' or laissez-faire
liberalism which was more dominant at the time), a strong case can be
made for seeing him as a prototype 'green' social theorist (Stephens,
1996; de-Shalit, 1997; Wissenburg, 1998; de Geus, 1999).

It is in Mill's eloquent defence of the 'stationary state' in which wealth,
capital and population are held as constant as possible, that is, a non-
growing economy, that the first outline of a 'green' or sustainable society
and view of the 'good life' can be found. To those who suggest that a
non-growing economy would be a social disaster, Mill replies:

> It is scarcely necessary to remark that a stationary condition of capital and
> population implies no stationary state of human improvement. There
> would be as much scope as ever for all kinds of mental cultures, and
> moral and social progress; as much room for improving the Art of Living,

and much more likelihood of its being improved when minds are ceased
to be engrossed by the art of getting on.

(Mill, 1900: 455)

Indeed, Mill suggests that in such a non-growing or 'steady-state
economy' (Daly, 1973), technological progress could and should be used
to lessen the work that has to be done, rather than simply be used to
employ less labour and thus enhance profits. This theme, which is at the
heart of green political economy (discussed later), again makes Mill
something of a green visionary. On the subject of technological
improvements, Mill expresses the view that in the stationary state,

> Even the industrial arts might be as earnestly and as successfully
> cultivated, with this sole difference, that instead of serving no purpose but
> the increase in wealth, industrial improvements would produce their
> legitimate effect, that of abridging labour. Hitherto, it is questionable if all
> the mechanical inventions yet made have lightened the day's toil of any
> human being. They have enabled a greater population to live the same life
> of drudgery and impoverishment, and an increased number of
> manufacturers and others to make fortunes.

(1990: 455)

Other 'green' aspects of Mill's thought is his defence of anti-cruelty to
animals legislation (Mill, 1900: 578), something which had a long
tradition in English **Utilitarian** social thought. Jeremy Bentham
(1748–1832), one of the founders of English Utilitarianism, for example,
was one of the first to give a philosophical justification for limiting
cruelty to sentient animals. According to him,

> The day may come when the rest of the animal creation may acquire
> those rights which could have never been witholden from them but by the
> hand of tyranny . . . The question is not, Can they *reason*? nor Can they
> *talk*? but Can they *suffer*?

(1823/1970: 311)

Utilitarianism, in basing its social and moral arguments on considerations
of pleasure, pain, welfare and happiness, was perhaps the first modern
social theory to be applicable beyond the species barrier. If one was
concerned with creating the 'greatest happiness of the greatest number'
(one of the common expressions of Utilitarian thought), then to restrict
one's concerns to humans only would be arbitrary and unjustifiable, since
like humans, animals are sentient, can feel pleasure and pain, experience
happiness and sadness and so on. This same Utilitarian argument since
first articulated by Bentham, then Mill, and also put into practice as the
animating principle of a popular 'animal welfare' movement of the late

nineteenth century, in which figures such as Henry Salt were prominent theorists and activists, can still be found today in the arguments of philosophers such as Peter Singer (1990) for 'animal liberation', and within the politics of the animal liberation movement.

This is not to say that there are other aspects of Mill's social, political and economic thought which would be at odds with green thinking. For example, though the general aim of Mill's remarks on 'The Stationary State' would be welcomed by many greens, his seemingly anthropocentric tone and language, and his somewhat over-dependence on population control, would alienate some of them. In his closing comments, Mill notes that, 'Only when, in addition to just institutions, the increase of mankind shall be under the deliberate guidance of judicious foresight, can *the conquests made from the power of nature* by the intellect and the energy of scientific discoverers, become the common property of the species, and the means of improving and elevating the universal lot' (1900: 455; emphasis added). Yet, despite (and indeed perhaps because of, as discussed later) the anthropocentric tone, the instrumental view of nature, and the almost exclusive concern with human progress and well-being, Mill can still be counted as one of the first green social theorists.

## Conclusion

It is with the Enlightenment or 'modernity', that we can see the first engagements between social theory and the environment, both as a precondition for society and when translated as 'nature' or 'natural' how it was put to a variety of ideological uses throughout the late eighteenth and nineteenth centuries. With the advent of modernity, 'nature' and the natural environment were, according to Weber, 'disenchanted', that is, the natural world was reduced to a set of 'natural resources' or 'raw materials' for human productive use.

Starting with Rousseau, some social theorists questioned the 'artificiality' of the modern world, and yearned for a return to a simpler, more 'natural' form of social order. This went against the prevailing view that life in the 'state of nature' was in the infamous words of Hobbes, 'nasty, brutish and short'.

While for some, human nature was 'naturally' competitive and indeed nature had been shown scientifically to be a struggle for survival (classical liberalism), for others, nature was a harmonious web of relations in which cooperation and mutual aid were just as important for evolution and progress (anarchism).

Within dominant nineteenth-century schools of social theory, such as liberalism and socialism, we can discern a common view of 'social progress'. Progress and 'development' were understood as a linear historical path from pre-industrial society to the complex and more advanced stage of the industrial socio-economic order. In many ways this particular linear view of human historical development is an onward and ever upward progression from ignorance, poverty, squalor and backwardness. While socialism criticised the capitalist character of industrial society, it did not criticise the industrialisation process in general or how the latter meant that the nonhuman world was 'dominated' and exploited, and viewed and used primarily as a means to human ends. While the other dominant nineteenth-century social theory, liberalism, generally agreed with socialism on the relationship between society and nature and the pursuit of material progress, J.S. Mill stands out against this dominant shared view of industrial development. As well as his defence of the extension of moral concern to nonhuman animals, his views on the desirability of a 'stationary state' mark him out as an early green social theorist.

## Summary points

- In social theory from the nineteenth to the twentieth century, an appeal to the 'naturalness' of principles or view of social relations has often been seen as indicating that such relations are 'given' that is, cannot be changed by human will.

- The idea of the 'state of nature' was the dominant way in which pre-Enlightenment social theory conceived of the environment, and it was regarded, by thinkers like Hobbes and Locke as inferior to civilised society.

- Rousseau went against this negative view of the environment and the state of nature with his ideas of the 'noble savage' and critique of the artificiality of civilisation.

- From 'Social Darwinism' to 'ecoanarchism' there have been ideological uses of the nonhuman world to justify, persuade and legitimate particular forms of social order.

- An early precursor of 'green' social thought is the anarchist reading of nature in which cooperation was seen as just as important as competition for natural, and thus social, evolution and development.

- Within dominant schools of social theory, such as liberalism and socialism, there was a common understanding of 'social progress'. Progress and 'development' were understood as a linear historical path from pre-industrial society to the complex and more advanced stage of the industrial socio-economic order.

- Marxist social theory criticised the capitalist character of industrial society, but not the industrial stage of social historical development. It accepted the necessity and desirability of 'subjugating' nature to humanity, or exploiting the natural environment so as to end human exploitation under capitalism.

- Mill and his brand of 'developmental' liberalism, to be distinguished from 'classical' or libertarian liberalism, anticipated many of the main concerns and aims of green social and political theory, particularly his defence of the 'stationary state' or non-growing 'steady-state economy'.

# Further reading

For a general overview of some of the theorists and theories outlined here (and others not discussed) see Derek Wall's *Green History* (London: Routledge, 1994), and Parts II and III of Marius de Geus's *Ecological Utopias: Envisioning the Sustainable Society* (Utrecht: International Books, 1999).

On pre-Enlightenment social thought and nature, see David Pepper, *The Roots of Modern Environmentalism* (London: Croom Helm, 1984). For a discussion of Hobbes and Locke see Keekok Lee, *Social Philosophy and Ecological Scarcity* (London: Routledge, 1989). Rousseau's thought and the environment are discussed in Shane Phelan's article (1992) 'Intimate Distance: The Dislocation of Nature in Modernity', *The Western Political Quarterly*, 45: 2. On the Enlightenment and environmental concerns see Alasdair Clayre (ed.), *Nature and industrialization* (Oxford: Oxford University Press, 1977).

The relationship between anarchist social theory and the environment is discussed in Murray Bookchin, *The Ecology of Freedom: The Emergence and Dissolution of Hierarchy* (Montreal and New York: Black Rose Books, rev. edn 1991); and Marius de-Geus *Ecological Utopias: Envisioning the Sustainable Society* (Utrecht: International Books, 1999).

For a discussion of Darwin and Malthus, see Donald Worster, *Nature's Economy* (Cambridge: Cambridge University Press, 2nd edn 1994).

On Marx, Darwin and environmental issues see Peter Dickens, *Society and Nature: Towards a Green Social Theory* (Hemel Hempstead: Harvester Wheatsheaf, 1992). On Marx and Engels on environmental issues see Henry

Parsons (ed.), *Marxism and Ecology* (Wesport, CT: Greenwood Press, 1977). For an assessment of the historical and contemporary relationship between Marxism and the environment see John Barry, 'Marxism and Ecology', in A. Gamble *et al.* (eds), *Marxism and Social Science* (London: Macmillan, 1999). Also see David Pepper, *Eco-Socialism: From Deep Ecology to Social Justice* (London: Routledge, 1993), Ted Benton, *Natural Relations: Ecology, Animals and Social Justice* (London: Verso, 1993) and Ted Benton (ed.), *The Greening of Marxism* (London: Guildford Press, 1996).

On liberalism, J.S. Mill and environmental concerns see Marcel Wissenburg, *The Free and the Green Society: Green Liberalism* (London: UCL Press, 1988) and Piers Stephens, 'Plural Pluralisms: Towards a More Liberal Green Political Theory', in Iain Hampsher-Monk and Jeffrey Stanyer (eds) *Contemporary Political Studies 1996* (Belfast: Political Studies Association, 1996).

**Twentieth-century social theory and the nonhuman world**

## Introduction

Following on from the last chapter, this chapter looks at some of the ways in which social theorising this century has dealt with, conceptualised and reacted both to the nonhuman world and to the relations between it and the human social world. Of particular interest are the ways in which the relationship between society and environment is theorised against the backdrop of the continuing legacy of the Enlightenment, and reflections on the development of modern, urban industrialised societies. This theme is particularly evident in critical theory and the work of two leading contemporary social theorists Jürgen Habermas and Anthony Giddens. Another emerging theme within twentieth century social theory is the connection between social theorising about external nature and internal or human nature.

## Classical sociology and the environment

Classical sociology, the 'science of society' has little to say about the environment or the environmental basis of human society. Where the nonhuman world did appear in classical sociology it was usually as one half of the organising dualism: society/nature, and/or as that which humans have historically overcome in their evolution from the Stone Age to the modern industrial age. As Goldblatt puts it:

> The primary ecological issue for classical social theory was not the origins of contemporary environmental degradation, but how premodern societies had been held in check by their natural environments, and how it was that modern societies had come to transcend those limits or had separated themselves in some sense from their 'natural' origins.
>
> (1996: 4)

In this sense, the social theory of such seminal thinkers as Max Weber, Emile Durkheim, Georg Simmel, Vilfredo Pareto and other founding fathers of sociology, while developing increasing complex and theoretically and empirically informed (if not always persuasive) explanations for social phenomena, had not progressed much beyond the eighteenth-century concern of social theory with exploring and charting the unilinear progress of society from 'rude' or 'primitive' to 'modern' or 'advanced' forms.

At the same time, there were disciplinary or knowledge-based reasons we can point to in order to understand the lack of concern within classical social thought with the relationship between society and its environment. Benton, commenting on the early development of sociology, suggests that:

> the conceptual structure or the 'disciplinary matrix' by which sociology came to define itself, especially in relation to potentially competing disciplines such as biology and psychology, effectively excluded or forced to the margins of the discipline such questions about the relations between society and its 'natural' or 'material' substrate.
>
> (1994: 29)

This observation concerning sociology can be expanded to classical social thought in general, as outlined in the previous chapter, was simply a continuation of the dominant approach to environmental considerations within nineteenth-century social thought.

## Freud, human nature and the war against nature

While an unlikely contributor to the engagement of social theory and the environment, Sigmund Freud (1856–1939), the modern founder of psychoanalysis, did have something to say on the issue. In *Civilization and its Discontents*, Freud argues that subduing external nature can be seen as part of the same project of humans imposing their wills on unruly 'nature'. The positive benefits that could be derived for humanity in uniting in a battle against nature included an unprecedented sense of collective solidarity, and a willed desire to eradicate nature-imposed problems (diseases, pain). In other words, if humanity were to regard nature as its 'common enemy', then it would have a strong reason to unite as a community in solidarity to 'fight' and 'subdue' nature. This aim echoed the early call of Francis Bacon for humanity, via scientific knowledge, to dominate nature. As Freud put it, his ideal for humanity involves 'combining with the rest of the human community and taking up the attack on nature, thus forcing it to obey human will, under the guidance of science' (quoted in Passmore, 1980: 23).

It is important to note that when Freud, like many later social theorists, spoke of 'nature', he meant both 'external nature', i.e. the natural environment, and 'internal nature', i.e. 'human nature'. In Freud's case, a central part of his theory concerned how civilisation, advancement and social progress depend not simply on the exploitation of external nature, but also required the 'controlling' of some potentially disruptive aspects of our 'internal nature'.

According to Freud, 'Our civilization is, generally speaking, founded on the suppression of instincts' (1950: 82). Hence the idea, developed later by critical theorists such as Marcuse (discussed below), that civilisation was 'against nature' in two senses. On the one hand, human civilisation was human, the product of human not nature's agency, and was created in opposition to the natural order (recall the binary opposition of 'culture/nature' in Chapter 2). On the other, the creation of a human social order, a civilised and secure social order, demanded the repression (or at least the sublimation) of human instincts, which were potentially wild, unruly and destructive.

In Freud's view of the relationship between the social and the natural order, we have not just another underlining of the radical separation of 'nature' and 'society', but the continuation of related theme which goes back to Hobbes. Like Hobbes, Freud was concerned that the 'natural impulses' of humans needed to be 'ordered' or 'contained' so that a human social order could emerge from natural disorder. Civilisation was not just threatened by 'nature' but also by 'human nature', the instinctual drives and modes of acting of humans as a species of animal. Hence the need to repress these potentially disruptive instincts within human social relations. And in arguing for humanity to unite in a 'war against nature' Freud was suggesting that these instincts were best channelled into subduing external nature. As Marcuse puts it: for Freud 'the diversion of destructiveness from the ego to the external world secured the growth of civilization. However, extroverted destruction remains destruction: its objects are in most cases actually and violently assailed, deprived of their form, and reconstructed only after partial destruction . . . Nature is literally "violated"' (1955: 78–9).

Thus Freud presented us with a picture of modern society ('modernity') in which the latter is premised on the double repression of 'internal' nature (transforming 'rude human nature' into civilised, ordered codes of conduct and manners) and the domination of external nature. One necessarily entailed the other. In calling for a united humanity to combat and control nature, Freud was simply suggesting a public, collective form of what modern industrial civilisation demanded at the individual, psychological level, an antagonistic and aggressive disposition and attitude towards both internal and external nature. This insight, not limited to Freud's work, set the scene for later social theorising about nature, particularly existentialism and critical theory.

## Existentialism and the 'meaningless' earth

While not strictly speaking a form of society theory, as opposed to a particular philosophy, existentialism is worthy of a brief discussion because it has an interesting slant on the relationship between humans and nature, and also lays the ground for much of recent social theorising about the environment.

Existentialism, associated with thinkers such as Jean-Paul Sartre, Martin Heidegger, Maurice Merleau-Ponty and Simone de Beauvoir, is an anti-rationalist philosophy of existence, which flourished from the 1930s to 1960s, though its modern origins go back to the Danish philosopher Søren Kierkegaard (1813– 55).

One of the distinctive ideas of existentialism is that unlike pre-modern and Christian thought, the existentialist position holds that humans are simply 'thrown into' a meaningless world. Whereas for both pre-modern, Christian and other forms of social theory, the natural world was basically our 'home', despite periodic problems between it and humans, there is no such comforting view within existentialism. Haught highlights how existential 'homelessness' can both explain the attractiveness of anthropocentrism and also account for environmental problems. For him,

> It appears quite likely that the origins of our environmental crisis lie, in part, in a deeply entrenched suspicion by humans that the cosmos is not really our home. The feeling of cosmic homelessness is, to a great extent, apparently 'religious' in origin . . . Exaggerating our own importance may be an understandable reaction to the prior conviction that we are exiles from any value-bestowing universe . . . Anthropocentrism is our way of responding to the feeling of not really belonging to the earth and the cosmos.
>
> (1994: 27–8)

Hence existentialism may be viewed as a consequence, how the separation of humans from nature (which as shown in Chapter 2 has a long tradition in Western thought) can easily lead to this existential sense of 'homelessness', isolation and alienation with regard to our relation to the environment.

This human separation and alienation from the natural world is the modern 'human condition'. Modernity and modern society have created a meaningless world which is indifferent to us and our fate. Existentialism is a very anthropocentric philosophical outlook in that its focus is on the meaning of human life, within society, with little reference to the relationship between society and environment. Like Nietzsche, existentialism takes the 'death of God' as a 'given' of the modern secular world in which Christian and other religious views are no longer culturally dominant and philosophically bankrupt in terms of telling us who we are and where we are going.

On the whole existentialism holds that our attitude towards the natural environment has to be an instrumental one, one in which the environment is a passive 'object' rather than an active 'subject'. In many respects, existentialism represents an extreme view of the consequences of the 'disenchantment of nature' in modernity. This is a theme taken up by one of the most influential schools of post-war social theory, the Frankfurt School of critical theory, which we turn to next.

# The Frankfurt School: critical theory, nature and the problems of modernity

Critical theory based in the Frankfurt Institute for Social Research (hence its later name, the Frankfurt School) is a body of neo-Marxist social theory. The distinctive features of critical theory were its critical analysis of the Enlightenment and the dominant ways of acting and thinking associated with it, and its linking social theory to social criticism of the prevailing 'modern' social order (both in the liberal-capitalist West and authoritarian-communist East). It included thinkers such as Max Horkheimer (1895–1973), Theodore Adorno (1903–69), Herbert Marcuse (1898–1979), Walter Benjamin (1892–1940), and Jürgen Habermas (1929– ) (discussed in more detail below).

Max Horkheimer and Theodore Adorno's famous work, *Dialectic of Enlightenment*, was an examination and explication of the 'dark' side of modernity, the costs and dangers of advanced technological, industrial society and its dominant modes of thought and behaviours. The danger they point out is quite stark. For them, 'The fully enlightened world radiates disaster triumphant', a sentiment with which many radical greens would agree in terms of the local and global environmental degradation and destruction modern industrial societies have caused. As Vogel puts it, 'the Frankfurt school's critique of contemporary society was offered up in a certain sense "in nature's name" – both that of the damaged inner nature of humans stuck in the fatal dialectic of enlightenment and an outer nature robbed of all qualities save those that render it amenable to human use' (1997: 175). As a result of the Enlightenment, the only value the natural environment can possess is instrumental value, that is, the natural world possesses value in so far as it is useful for human purposes or ends.

The 'disenchantment of nature' (the cultural transformation of nature from a morally significant realm with its own intrinsic value, to being viewed solely as a set of resources for human use and enjoyment, discussed in the last chapter), as one of the main consequences of modernity, was something not just to be regretted (the Romantic reaction) but also dangerous for both human society and the nonhuman world. In particular, the increasing rationalisation which was central to the successful technical manipulation of external nature had a tendency to 'spill over' into other spheres of human life in which they were not appropriate and were dangerous. The basic problem was this: the instrumental use of nature developed institutions, modes of thinking and acting which were then 'transferred' illegitimately to human social and personal relations. The domination and exploitation of the natural environment leads to the domination and exploitation of humans. Enlightenment-derived institutions as modes of rationality suited to human-nature exchanges, contained the possibility of being used in human social relations where they were dangerous and harmful. As Horkheimer and Adorno put it, 'Men have become so utterly estranged from one another and from nature that all they know is what they need each other for and the harm they do to each other' (1973: 253).

## Marcuse and the 'liberation of nature'

In *Eros and Civilisation*, his study of Freud, Marcuse describes the relationship between the 'death instinct' and the exploitation of the external world. For Marcuse,

> the entire progress of civilization is rendered possible only by the transformation and utilization of the death instinct or its derivatives. The diversion of primary destructiveness from the ego to the external world feeds technological progress . . . In this transformation, the death instinct is brought into the service of Eros; the aggressive impulses provide energy for the continuous alteration, mastery, and exploitation of nature to the advantage of mankind. *In attacking, splitting, changing, pulverizing things and animals (and, periodically, also men), man extends his dominion over the world and advances to ever richer stages of civilization.*
>
> (1955: 47; emphasis added)

Later, in a critique of the Enlightenment project, Marcuse asserts that 'The ego which undertook the rational transformation of the human and natural environment revealed itself as an essentially aggressive, offensive subject, whose thoughts and actions were designed for mastering objects. It was a subject *against* an object' (1955: 99). In other words, the human subject (the modern self) in order to dominate the external environment to produce the industrial social order had to see that external world as a passive object for human manipulation and control. In so doing, as Marcuse points out, this attitude towards nature (including human nature) revealed an 'aggressive, offensive' human self and by extension, an aggressive, offensive modern social order (capitalism). As he goes on to explain, 'Nature (its own as well as the external world) were "given" to the ego as something that had to be fought, conquered, and even violated' (1955: 99). Human social development and progress under modern industrial capitalism required, as Freud echoing a theme of social theory going back to Hobbes and Rousseau, a warlike attitude towards the natural environment and a will to control the potentially destabilising instincts of internal (human) nature.

However, the 'reconciliation with nature' that Marcuse advocated did not imply going back to some 'pre-modern' or 'organic' relation between humans and nature. On this point Marcuse, following Horkheimer and Adorno, is clear. For him, the '"liberation of nature" cannot mean returning to a pre-technological stage, but advancing to the use of the achievements of technological civilization for freeing man and nature from the destructive abuse of science and technology in the service of exploitation' (1972: 60). He argues instead for a 'liberating' domination of nature as opposed to the 'repressive' one typical of modern industrial societies, and suggests that 'no free society is imaginable which does not . . . make a concerted effort to reduce consistently the suffering which man imposes the natural world' (1972: 68). This 'liberating domination' means the 'civilising' or humanising of nature through such liberating physical (and cultural) transformations of nature from 'wilderness' or 'raw nature' into parks, gardens, farmland, landscapes and reservations. Here we find echoes of the Christian 'perfection of nature' idea discussed in Chapter 2.

A central part of reducing this unnecessary suffering for Marcuse requires a transformation in the idea and practice of 'development' in the advanced capitalist nations, basically a shift from a quantitative, consumption-orientated view to a more aesthetic-qualitative one. Anticipating a central green or ecological argument, he suggested that 'the sheer quantity of goods, services, work and recreation in the overdeveloped countries which effectuates this containment. Consequently, qualitative change seems to presuppose a quantitative change in the advanced standard of living, *namely, reduction of overdevelopment*' (1964: 242; emphasis added). And like the green argument for reducing 'overdevelopment', Marcuse suggested this for both ecological reasons (decreasing environmental degradation) and social/emancipatory ones (such a transformation would lead to a less oppressive, bureaucratic, work-orientated, consumption-based society).

Towards the end of his life he saw in the emerging ecological movements in Europe and North America a more promising anti-capitalist social movement than the Marxist class struggle, which expressed the radical 'emancipatory' goals he endorsed, and the creation of a 'post-capitalist' society. As he put it,

> The ecology movement reveals itself in the last analysis as a political and psychological movement of liberation. It is political because it confronts the concentrated power of big capital, whose vital interests the movement threatens. It is psychological because . . . the pacification of external nature, the protection of the life-environment, will also pacify nature within men and women A successful environmentalism will, within individuals, subordinate destructive energy to erotic energy.
>
> (1992: 36)

However, since Marcuse critical theory has not made ecology, the nonhuman world or social-environmental relations central themes in its critical analysis of modern societies. As Whitebrook has pointed out, 'critical theory, which aspires to provide a comprehensive theory of the crisis of modernity, has little to say about one of its most decisive features, ecology' (1996: 286). Partly for an assessment of this claim we turn to perhaps the most important theorist of contemporary critical social theory, Jürgen Habermas, next.

## Jürgen Habermas and the problem of nature in modernity

In the main Habermas has sought to show that the only relation we have with the natural environment is an instrumental one, governed by productive, prudential and technical concerns about how best we can exploit it. His concern, relating to what he sees as one of the dangers of modernity, is to prevent human social relations (or what he calls communicative concerns) from being reduced to instrumental norms which are appropriate to the sphere of human technological manipulation of the natural world. That is, he does not want how we treat and view each other to be the same as how we treat and view the nonhuman environment.

Habermas's basic position is that an instrumental, technical or manipulative attitude towards the external world is something that is simply a 'given', a 'brute fact' of the particular character or nature of the human species. For Habermas humans are genetically and evolutionarily disposed towards having this view of the natural environment, or, as Vogel puts it, we have 'hard-wired interests' (1997: 183) in nature of an instrumental character, and these cannot be changed. The evolutionary character of Habermas's thought, which is in keeping with classical social theory, is central to understanding his position. However, unlike classical social theory, according to Dickens (1992), Habermas is at pains not to be seen to promote a simplistic, unilinear view of social development. However, what Habermas does share with classical social theory is the Enlightenment belief that the progress and development of human society is premised on the exploitation and instrumental use of the natural environment.

## The moral status of the nonhuman world

However, Habermas seems uneasy with the fact that his theory 'seems to preclude as irrational a non-objectivistic relation to nature' (1982: 241). That is, it seems that Habermas must dismiss as 'irrational' a non-instrumental view or valuation of the natural environment. It seems that this instrumental-technical attitude to nature is something we must simply accept as the 'price' to be paid for the advantages of the modern social world. Habermas's view of social-environmental affairs is heavily influenced by his theory of human knowledge in which the external environment is for him 'constituted' by the type of knowledge appropriate to its study, namely the natural sciences and the instrumental view and attitude they have towards nature. He does not seem to deny non-instrumental attitudes to nature, but rather maintains that an instrumental attitude is superior in terms of human manipulative control over nature. As he puts it, 'while we can indeed adopt a performative attitude to external nature, enter into communicative relations with it, have aesthetic experience and feelings analogous to morality with respect to it, there is for this domain of reality only one *theoretically fruitful* attitude, namely the objectivating attitude of the natural-scientific, experimenting observer' (1982: 243–4; emphasis in original). At the same time, Habermas is concerned that the alternative to this instrumental-technical attitude to nature (based on its 'disenchantment') is a dangerous strategy of 're-enchanting' nature via an appeal to some mystical or spiritual worldview. For him, 'The phenomena that are exemplary for a moral-practical, a "fraternal", relation to nature are most unclear, if one does not want to have recourse here as well to mystically inspired philosophies of nature, or to taboos (e.g. vegetarian restrictions), to anthropomorphising treatment of house-pets and the like' (1982: 244–5). And the problem with such strategies for re-enchanting nature is that they do not lend themselves to democratic or discursive modes of communication, as well as entailing a breakdown in rational scientific discourse. Mystical attitudes are by definition beyond rational explication and justification, and for Habermas this is precisely what makes them invalid: such forms of thinking and acting constitute potentially anti-democratic forms of authority.

Yet while he rejects attempts to 're-enchant' the environment, he cannot as easily dispel what can only be described as an unease with the implications of his theory for the negative affects of human action on the nonhuman world. Thus, for example, he notes that 'The impulse to provide assistance to wounded and debased creatures, to have solidarity with them, the compassion for their torments, abhorrence of the naked instrumentalisation of nature for purposes that are ours but not its, in short the intuitions which ethics of compassion place with undeniable right in the foreground, cannot be anthropocentrically blended out' (1982: 245).

An almost exact repetition of this line can be observed in a later aspect of his thought, that concerned with 'discourse ethics', where his ambiguity about the claim that our relations with the nonhuman world are normative is evident (if not fully resolved). Habermas asks,

> How does discourse ethics, which is limited to subjects capable of speech
> and action, respond to the fact that mute creatures are also vulnerable?
> Compassion for tortured animals and the pain caused by the destruction
> of biotopes are surely manifestations of moral intuitions that cannot be
> fully satisfied by the collective narcissism of what in the final analysis is
> an anthropocentric way of looking at things.
>
> (quoted in Outhwaite, 1996: 200)

Unfortunately, no satisfactory answer from within his own frame of reference is forthcoming; and on the whole, while the place of nature and social-environmental relations does constitute a gap in Habermas's otherwise impressive and massive social theory, he does not seem unduly bothered by this. Some of his followers on the other hand have taken up this task on his behalf, so to speak, and it is to some of these that we turn next.

## Knowledge, nature and society: towards reconciliation?

According to Vogel, one of the main roots of Habermas's 'unease' with the place of nature within his theory lies in his theory of knowledge.

Habermas maintains a dualism between society and nature and the types of knowledge appropriate to these two spheres. He separates the 'natural' from the 'social' sciences, not least by stating that the former is 'positive' in the sense of yielding objective, value-free knowledge, while the latter is 'normative', that is, yields knowledge about society and social relations which depend on values and normative principles. According to Vogel, 'The cost of this inoculation of natural science from the normative ... is precisely the impossibility of conceptualizing nature itself as possessing moral worth, or of recognizing a moral dimension to the way in which we interact with it' (1997: 180). This is simply another way of stressing the 'price' to be paid for the 'disenchantment of nature' as a necessary precondition for increased human technological power and control over nature, and the material and other benefits that follows from this.

Vogel contends that Habermas has an unduly narrow view of scientific knowledge, and contends that there are other 'theoretically fruitful' (to use Habermas's own terms) ways of studying, conceiving and interacting with the natural environment. For him:

> there is no single 'theoretically fruitful' approach to nature, but rather the question of what is fruitful turns out itself to be answerable in socially and historically varying ways, which means that alternative approaches are certainly imaginable. The view of nature as mere matter for instrumental manipulation criticized by Marcuse and the earlier Frankfurt School is not, as Habermas tried to argue, built into the structure of the species or of 'work', but rather – as they had originally asserted – is associated with a particular social order and a particular historical epoch. A new kind of society, then, might well involve a new science, and with it a new nature' as well.
>
> (Vogel, 1997: 187).

In this way Vogel is presenting a form of the social construction of nature in that different social orders may have different forms of science which in turn will reveal or construct new forms of natural environment for study and contemplation. Thus Vogel contends that a 'reconciliation of humans and nature', one of the aims of the early Frankfurt School (and of contemporary deep ecologists and others in the radical ecology movement discussed in Chapter 9) is possible, but only if what he calls the 'sociality of nature' is explicitly acknowledged. For him, the natural is always the social in two senses, 'first, because we perceive and experience it, study and dream about it, in terms that are from the beginning social through and through, *but second also because the objects and landscapes through which we experience it are always themselves – when closely examined – in part the product of earlier social practices'* (1997: 186; emphasis added). While the first can be accepted, in the sense that 'the environment' as a concept is constructed by humans, the second claim that the environment is, in part, always also physically produced by humans is highly contentious.

The reconciliation he outlines is not based on re-enchanting nature, nor premised on recognising the 'Otherness' of nature (that is, how radically different or alien nature is, from a human point of view, discussed later in Chapter 7) or attempting to see nonhuman entities as potential participants in Habermassian 'discourse ethics' as Dryzek (1990), for example, attempts. According to Vogel, 'What is wrong with the way nature appears to us (and our natural science) today is that it seems utterly independent of us and even opposed to us. Its sociality is hidden: we see it as separate from us, as dangerous (and impossibly complex, as always taking its revenge), and fail to see that rather it is always something in which we are deeply and actively enmeshed' (1997: 188). It is precisely because nature appears as 'other' as 'alien' that we feel separate from it and consequently fear it, and this leads to a desire to control and dominate it. Vogel's strategy is to claim that this seemingly 'alien', strange external nature is in fact always already social in that it is partly the product of (past and present) human transformative practices. As he puts it, 'Only a nature viewed as so separate from us would be something we would feel the (frightened) need to "dominate" or the (equally frightened) need to "preserve". In either case it appears as alien, something to be either overcome or propriated – not simply as part of the (social) world that we both inhabit and continuously transform' (1997: 188). If we see that nature is already social, something which humans have helped produce, this brings home to us that it is not alien, not 'other' and thus lays the basis for harmony between humanity and nature. Vogel's philosophical attempt at reconciliation is based on the (Marxist) idea that we can know that which we make. Hence if 'nature' is, at least in part, 'socially constructed' in the ways Vogel maintains, we can know it and thus nature is no longer alien or 'other'.

While there is much merit in Vogel's Habermassian 'environmental ethics', in particular his argument which shows the failings of Habermas while also pointing to ways in which Habermas's theory can take on board environmental considerations, there are some problems with it. The largest problem is that Vogel's position cannot extend beyond those parts of the natural environment which are resolutely not, even in part, the product of (past or present) human social practices. Thus while Vogel's argument is generally persuasive in respect to parts of the natural environment like agricultural landscapes, parks, gardens, and other natural systems which are either wholly or partly the product of human transformation, it is less persuasive when it comes to 'wilderness' areas and other aspects of the natural world, such as global hydrological and carbon cycles, the ozone layer, etc., which are clearly not the product of human social practices. Thus the second of Vogel's claims about the 'social' construction of the environment is, at least for these parts of the natural environment, extremely doubtful. Since we do not make these categories of natural entities, processes and systems, we cannot 'know' them in the way demanded by Vogel's reconciliation thesis. And in this respect, an appeal to the 'otherness' of nature may be unavoidable, though of course this does not mean a sense of 'alienation' from nature (O'Neill, 1993), or necessitate a mystical re-enchantment of nature (Barry, 1993) to overcome that alienation.

## Environmental politics and the defence of the 'lifeworld'

Habermas, like Marcuse before him, sees the rise of environmental politics (as a 'new social movement'), expressing a concern for the care and protection of the natural environment, as significant. Environmental politics are significant according to Habermas because they mean a new politics beyond 'left and right'. As he puts it, 'the[se] new conflicts are not sparked by *problems of distribution*, but concern the *grammar of forms of life*' (1981: 33). By this he means that green politics and concern for the preservation of the natural environment mark a new form of politics, one that is not focused on the distribution of the economic pie, jobs, consumption, income, wealth and so on, which are the main issues in left–right politics. Rather, environmental politics, along with other new social movements, like gay rights and feminism, seek to challenge and change the existing institutional and moral order, a major part of which involves bringing the question of identity and lifestyle to the centre stage of political struggle.

Environmental politics seek, in part, to protect what Habermas calls the 'lifeworld', the world of everyday social interaction and realm of moral action from the 'systems world', the world of bureaucratic, state administration and the capitalist market. However, whereas for Marcuse, environmental politics is radical and emancipatory, for Habermas they are 'defensive', a reaction against the technological domination of the natural world in modern advanced societies.

One of Habermas's main concerns is with preventing what he calls the 'scientisation of politics': that is, ensuring that political discourse is not reduced to a 'technical' one, so that how humans treat each other (through political action and institutions) does not resemble how humans treat nature. However, for Habermas environmental problems are essentially 'technical' problems not moral ones, which goes against how many of those concerned about the natural environment (and not just greens) feel about environmental degradation and destruction. As he puts it, 'The intervention of large-scale industry into ecological balances, the growing scarcity of non-renewable resources, as well as demographic developments present industrially developed societies with major problems; but these challenges are abstract at first and call for *technical and economic solutions*, which must in turn be globally planned and implemented by administrative means' (quoted in Outhwaite, 1996: 325). Thus, Habermas can be said to continue the pervasive dualism with Western social theory which insists on maintaining a sharp and morally significant separation between the social and the natural worlds. While not completely the case, it is at the same time not completely misleading to see Habermas as being more interested in the human 'lifeworld' than the 'natural world'.

## Anthony Giddens, globalisation and the environment

Like Habermas, Anthony Giddens's work is concerned with modernity and its effects, though unlike Habermas he has paid more attention to environmental issues, and the place of nature within his social theory. Giddens's engagement with environmental issues begins from an awareness of the lack of attention to ecological issues within sociology. As he puts it, 'Ecological concerns do not brook large in the traditions of thought incorporated into sociology, and it is not surprising that sociologists today find it hard to develop a systematic appraisal of them' (quoted in Cassell, 1993: 287).

Giddens's explicit concern with 'space and time' also means that he is more sensitive than other social theorists to ecological concerns, which clearly have spatial/geographical and temporal dimensions and import for human societies. This geographical dimension of Giddens's work is central to his theory of globalisation. Though the idea has undergone some alteration within Giddens's work, as Goldblatt (1996) points out, globalisation means the linking of the global and the local as a consequence of time–space distantiation. Simply put, globalisation refers to the historical processes which for the past four or five centuries have been connecting and bringing various parts of the world together into one system of cultural, political and above all, economic relations. While this process has been going on for some time, it has intensified since the Second World War. As a result of various developments politically, culturally, in communications and transport, and in the establishing of a global capitalist market, the creation of an international division of labour, the rise in prominence of transnational corporations, the world today is a smaller place. By this is meant that time and space have been compressed in what Giddens calls the 'late modern age'. As a result of globalisation, distant places on the globe are intertwined and interdependent in that they are tied into a series of shared and common relations, institutions and processes.

Giddens's sensitivity to how central environmental issues are to modern social inquiry can be seen in his analysis of globalisation where he writes that 'the diffusion of industrialism has created "one world" in a more negative and threatening sense . . . a world in which there are actual or potential ecological changes of a harmful sort that affect everyone on the planet' (1990: 76–7). The modern world is a global world in the sense that there is a link between the local and the global, as a result of space–time compression, such that changes in one part of the world can have potentially devastating effects on another part. While in Giddens's theory of globalisation these effects range from economic, financial, and cultural consequences (not all of which are necessarily bad), one of the most tangible experiences of globalisation are global ecological problems. In particular, pollution problems, global warming and climate change transcend territorial boundaries of nation-states and are 'global' in their scope (which is not to say they affect all places and people on the planet equally). Thus the spread of global ecological problems is a specific consequence of globalisation, the transmission and spread of industrial capitalism to the entire globe, the creation of a global market, and the development of various communication, political and cultural institutions and connections bringing distant societies together within a single globalised world system.

## The rise and meaning of environmental politics

According to Goldblatt (1996: 69–71) we can discern at least three different explanations of environmental politics within Giddens. The first is what we can call a 'conservative' view of environmentalism in that for him environmental movements are associated with recovery, the recovering of traditional ways of relating to the environment. Here the urban environment produced environmental movements born of the cultural alienation and spiritual vacuum of the modern urban landscape. A second explanation Giddens has for environmental politics is as a response to perceived ecological threats. On this account, environmental politics emerges as ecological problems and dilemmas become more obvious and discernible to people.

A third explanation is that environmental politics is a 'lifestyle politics' associated with new social movements. Here, Giddens is close to Habermas's view which associates green politics with non-distributional issues. For Giddens, green/environmental politics is concerned with how one should live and issues of personal identity, rather than the typical issues which dominate 'mainstream' (left–right) politics, such as income levels, employment and economic growth. As Goldblatt puts it, for Giddens, 'environmental politics is not simply the outcome of increasingly perceived environmental risk. It is also fuelled by an increasing demand for the remoralization of abstract systems of social organization that have ceased to be accountable in any meaningful way to those they affect' (1996: 71). Thus Giddens sees the increasing concern with environmental issues and the rise of environmental politics as developments which are explicitly moral, in raising moral questions about the modern social order, its institutions and principles.

## The urban environment and the 'town' versus the 'countryside'

According to Giddens,

> A critical theory alert to ecological issues cannot be limited to a concern with the exhaustion of the earth's resources . . . but has to investigate the value of a range of relations to nature that tend to be quashed by industrialism. In coming to terms with these we can hope not so much to 'rescue' nature as to explore possibilities of changing human relationships themselves. An understanding of the role of urbanism is essential to such

an exploration. The spread of urbanism of course separates human beings
from nature in the superficial sense that they live in built environments.
But modern urbanism profoundly affects the character of human day-to-
day social life, expressing some if the most important intersections of
capitalism and industrialism.

(quoted in Cassell, 1993: 329)

What is interesting about Giddens's writing on environmental issues is
the stress he lays on the need for social theory to engage with urban, built
environments. Not just as they constitute the lived, everyday
'environment' within which individuals are situated, but also because the
effects of urbanism have ramifications for how the 'natural environment'
is constituted, perceived and acted upon. Here, Giddens is in a tradition
within social theory in which urbanisation, cities, buildings and the
creation of human-made artificial spaces and places are significant
because they represent or express the difference between 'modern' and
'pre-modern' society.

According to Goldblatt, 'urbanism', for Giddens, is the mediator of new
modern relationship to the natural world. 'modern urbanism is the point
at which the culturally transformed experience of the natural world is
most acutely felt . . . capitalist urbanism is the physical site of the
wholesale transformation of the natural into wholly manufactured space'
(Goldblatt, 1996: 56). Thus the spread of urban life in modern societies
(and modernising societies) marks a profound cultural shift in how the
natural environment is experienced, viewed and valued.

Building on Giddens, we can say that one of the consequences of urban
modernity is the loss of a sense of being within the 'natural order'. Thus
while a modern as much as a pre-modern society is fundamentally
dependent upon the natural world (a point stressed by green social theory,
discussed in Chapter 9), living in an urban environment distances people
from this reality: the reality of the natural world and the material reality
of society's dependence upon it. As Goodin puts it, in discussing the
normative basis of green political theory, people wish to belong to a
whole that is larger than themselves (1992: 38), and for most of human
history up until the modern age, nature provided just such a stable,
permanent order. However, living in an urban environment, removed
from the material or ecological reality of human dependence upon the
environment, there is a marked tendency within modern urban societies
to assume that natural limits, constraints have been 'transcended'.
Urbanisation removes the natural environment from the everyday lives of
people, replacing it with an artificial, human-made one.

At the same time individuals are removed from the reality of modern productive interactions with the natural world, people do not know exactly how the food, energy and other commodities they consume are produced, where they are from, who produced them, under what conditions, etc. In this sense, modern urbanism complements another ecologically harmful feature of contemporary global capitalism: the increase in distance between production and consumption (Barry, 1999: ch. 6). Thus while globalisation compresses space and time, it can also by the same token increase distances between production and consumption.

Giddens's account of urbanism needs to be extended beyond a concern with 'manufactured space' to include how urbanism can remove the natural environment from centre stage, as the immediate environment of day-to-day life, diminish its role as the permanent backdrop for human action, push it backstage as it were, and create the impression that humans do not depend upon the natural world.

For Giddens, faced with the alienation and insecurity of the modern, mobile world, environmental politics represents an attempt to establish some substantive moral content, and normative security to people's lives. The reason for this has partly to do with the paradox of modernity in relation to the environment. This paradox is that the more society and the natural world are 'modernised', that is, brought under the regulation of the bureaucratic nation-state and the market, the greater the perception of and the higher the value placed on those parts of the natural world which have not been modernised or developed. Giddens's point is that, at least in part, the distinctly modern concern with preserving and protecting the environment, as well as having to do with the reality of environmental problems, has a deeply moral element. And it is this deeply moral element which makes green politics 'beyond left and right' (Giddens, 1994).

One the one hand, the increase in severity and public awareness of environmental problems are motivated largely by collective and individual interests in survival, that is they are protective and reactive measures against (human-caused) environmental dangers. A central consequence of this is a desire or need to maintain the natural environment in some state at which environmental problems for humans are solved or prevented. On the other hand, the moral content of environmental politics has to do with defending a particular *meaning* of the natural environment. While he does not develop his discussion of the environment in this way, it is compatible with Giddens's social theory to suggest that the particular meaning of the environment in question for environmental politics has something to do with the environment as a separate, independent, stable 'natural order' within which humans can find security and a meaningful order.

It is also related to particular understandings of collective identity, and tradition according to Giddens (1994: 206). While this particular meaning of the natural environment can be detected in many aspects of the environment, it is perhaps most obvious in the dominant meanings attached to the 'countryside' as opposed to the 'town' (Williams, 1973). As Rennie-Short points out, 'In the contrast with city, court and market, the countryside is seen as the last remnant of a golden age. The countryside is the nostalgic past, providing a glimpse of a simpler, purer age . . . The countryside has become the refuge from modernity' (1991: 31, 34). Hence the connection between environmental protection and tradition for Giddens. Indeed, in discussing Robert Goodin's (1992) argument for the value of nature resting its capacity to be a context larger than ourselves, Giddens notes 'To say that we need something "larger" than ourselves or more enduring than ourselves to give our lives purpose and meaning may be true, but this is plainly not equivalent to a definition of the "natural". It fits "tradition", in fact, better than it does "nature"' (1994: 206).

## Environment, tradition and identity

The relationship Giddens talks about between particular forms of collective identity, tradition and particular meanings attached to the 'natural environment' can be seen by looking at the case of the 'countryside' and how it has been used in recent debates about competing conceptions of 'English' national identity. It must be remembered that when speaking of the countryside, one is not strictly speaking referring to a purely 'natural' (in the sense of 'nonhuman' or 'non-transformed') environment. The countryside is a humanised environment in that it is not a naturally occurring natural ecosystem. Hedgerows, ploughed fields, drystone walls, all of which are central to the idea of countryside, are the product of present and past human transformation of the natural environment. In many ways the countryside is a 'text' which contains the 'trace' of previous human transformative activity. Preserving the countryside can thus be motivated by a desire to maintain the continuity with previous generations, to preserve a particular way of life, and a particular sense of collective identity. For example, there is an argument to suggest that it is a particular conception of 'Englishness' which can, in large part, explain recent movements to 'protect the countryside'. These movements ranging from the Countryside Alliance to the National Trust and Council for the Protection of Rural England, are motivated not simply by a concern to protect the countryside per se, but to protect a particular type of countryside and land-management system which is a constitutive aspect of a particular understanding of Englishness and 'tradition'. This idea of Englishness, in part, has to do with such claims as England as a 'green and pleasant' land, a land of sturdy yeomen farmers, village greens, a hierarchical but benign class system in which the aristocracy are the main landowners who 'take care' of their land-renting peasants, a society where fox-hunting is not simply a way of getting rid of a pest, but also a constitutive part of the country way of life, and so on.

Many of these arguments are explicit in the aims of the Countryside Alliance who saw in the defeated 1998 Parliamentary Bill to outlaw fox-hunting an attack on the country way of life by urban, liberal 'townies' who know nothing about the reality of rural life. Thus in this conflict between 'defenders' of fox-hunting and those opposed to it, we find a contemporary example of a long-standing relationship between 'country' and 'town'. Thus, while ostensibly about the practical and moral issues surrounding a particular practice (fox-hunting), it is clear that there is more to this conflict than this, which is not to downgrade the moral commitments and concerns for the welfare of foxes opponents have. In short, we can say that what is at issue is also a conflict between two different and potentially incompatible views of or meanings of the 'countryside'. The important point to note is that it is not the physical reality of 'natural environment' or the 'countryside' as such that is the object of analysis, but the 'meaning' of the countryside (which of course will affect how it is physically affected by human action).

The Countryside Alliance is defending a view of the country which stresses the idea of the rural way of life, and thus emphasises how the countryside is not simply a 'rustic landscape' but a *working environment*, and a system of land management which requires fox-hunting. They accuse opponents of fox-hunting of regarding the countryside as some pastoral landscape, a place of hedgerows, butterflies, copses, forests, babbling brooks, etc., in which the economic reality of country life is missing. From their perspective the urban, metropolitan opposition to fox-hunting on the grounds of its 'uncivilised', 'barbaric' character, is in fact the town misperceiving the countryside as the 'pastoral' in Rennie-Short's (1991) terms, and thus 'forcing' the rural way of life into something which it is not.

At the same time, there is a conflict over the relation between 'Englishness' and the countryside in this debate. Defenders of fox-hunting can justify it on the grounds that there is something quintessentially 'English' about fox-hunting. Oscar Wilde's famous definition of an Englishman fox-hunting as 'The unspeakable in pursuit of the inedible' stands with conservative views in which a true Englishman must 'ride to hounds' as examples of this connection between particular environmental practices (and thus particular meanings of the environment) and collective identity. Again Rennie-Short eloquently highlights this identity-forming and identity-affirming function of the countryside, 'In most countries the countryside has become the embodiment of the nation, idealized as the ideal middle landscape between the rough wilderness of nature and the smooth artificiality of the town, a combination of nature and culture which best represents the nation-state' (1991: 35).

Another example of this connection between the countryside and collective identity is the concern expressed over 'foreign' or 'non-indigenous' species of flora and fauna displacing indigenous species. The most striking example here are the debates over grey and red squirrels, and the worrying fact that the native red squirrel is being driven from its native habitat by the 'foreign' grey squirrel (and thus its numbers are falling). In this debate over the need to protect the native species from the foreign one, more than simple ecological or animal welfare concerns are at stake, just as was argued to be the case in relation to fox-hunting.

At the same time, there are other movements such as The Land Is Ours and to a lesser extent the Ramblers' Association, which articulate another alternative understanding of the countryside, this time with an emphasis on the right of every British citizen to have access to the land, both in terms of walking and enjoying the countryside (the Ramblers' Association), and the more radical aim of returning ownership and control rights of the land to the people. In such movements, particularly The Land Is Ours, one can see strong connections to past social struggles and environmental practices. For example, the aims of The Land Is Ours movement echo the demands of earlier radical English movements such as the Levellers and the Diggers to defend the 'commons' and the rights of access of the commoners. However, its core democratic aim of redressing the situation today in which 80 per cent of the land in England is owned and managed by 10 per cent of the population, also makes it distinctly 'modern' in aspiration. Though beyond the scope of our present discussion, one could suggest that in such contemporary movements there is an attempt to create a newer, less exclusive and monolithic sense of Englishness in relation to the countryside and particular social-environmental practices such as fox-hunting, which marks a decisive break with older conceptions of English national identity which were in large part based on a particular conception of the English countryside, the country way of life, and England as a 'green and pleasant land'.

Unfortunately, these issues are largely missing from Giddens's work on the place of the environment within social theory – largely, I suspect, because he is insufficiently sensitive to the important differences and conflicts between competing meanings of different parts of the environment. Thus while as Goldblatt notes, for Giddens, 'the conjunction of capitalism and industrialism is responsible for modern environmental degradation . . . whatever the precise causal origins of environmental degradation, the modern world heralds a more wholesale transformation of nature than human societies have been capable of before' (1996: 16–17), Giddens's theory is in many important respects limited. Together with a lack of attention to the full range of issues within the urban/rural or town/country divide, there are other limits to Giddens's social theory.

Relating to the distinction between 'transformed' and 'natural' environments, the idea of 'wilderness' is problematic for Giddens, just as it was for both Habermas and Vogel discussed earlier. An example of this is Giddens's statement that, 'In the industrialised sectors of the globe – and increasingly, elsewhere – human beings live in a created environment . . . *Not just the built environment of urban areas but most other landscapes as well become subject to human co-ordination and control*' (1990: 60; emphasis added). We can note that 'landscapes' are not co-extensive with the natural environment. In this sense Giddens has shifted, and in the process narrowed, the focus of the debate from an analysis of the full range of issues involved in relating the environment to social theory. He focuses on 'landscapes', which for him includes the urban landscape, or what has been termed 'blandscapes' by writers such as Porteous (1997), on account of the homogeneity, uniformity and lack of aesthetic content that is typical of most urban, built environments.

Giddens misses the essentially ecological character of the environment, and its importance for human societies, and substitutes it with an aesthetic-moral concern with the landscape/blandscape of the modern social world. This is not, let me stress, to deny the importance of this aesthetic-moral concern with producing pleasing, beautiful and enjoyable landscapes, but this concern must be placed within its ecological context, and the relationship between society and *environment* and not just society and *landscape*. Here, however, Giddens may be working with a similar notion to Marcuse's 'liberatory domination of nature', discussed above. For Giddens there is no necessary connection between human mastery and exploitation of nature and the destruction of the natural environment. As he puts it, 'Mastery over nature . . . can quite often mean caring for nature as much as treating it in a purely instrumental or indifferent fashion' (1994: 209).

Secondly, it is surely overstating the case to suggest that 'most other landscapes' human beings experience or come into contact with are (and ought to become?) transformed by human practices. While this may be true of the highly populated societies of Western Europe, whose indigenous, local environments have been intensively and extensively transformed by previous human activity for hundreds of years (and thus cannot be considered as 'wilderness' and is closer to a 'garden' in terms of the discussion earlier in Chapters 2 and 3), this is not the case with other societies around the world, where wilderness does exist. To the extent that Giddens relies on this idea of the environment as one transformed by 'human co-ordination and control' as the primary orientation around which to integrate the environment into his social theory, to that extent it is limited to the 'humanised environment' rather than the nonhumanised natural environment. This is confirmed by his statement that, 'All ecological debates today, therefore, are about managed nature' (1994: 211). This focus on 'managed nature' becomes even more problematic if we include natural environmental processes such as hydrological cycles, carbon and nitrogen fixing, and ecosystem functions, as part of the natural environment we are interested in. These processes are not, and could not be, subject to human manipulation and control.

However, despite these problems, Giddens has developed a coherent and challenging analysis of the place of the environment (both natural and urban) within social theory. He presents us with an account of the environment in modern society and social theory in which the manufactured or managed character of the natural and urban environments, within the context of globalisation, are given central stage. Without looking at urbanisation, any analysis of the emergence of green politics and a moral concern for the preservation of the natural world is deficient for Giddens. Green issues and politics, and the place of the environment in social theory, are distinctly 'modern' phenomena, and the need to be sensitive to the urban experience is the mark of how modern these concerns are. They arise, in large part, as a result of what he calls a paradox of the modern world, which is 'that nature has been embraced only at the point of its disappearance' (1994: 206).

## Conclusion

A central theme of twentieth-century social theory's engagement with the environment has focused on the costs (social and psychological as well as environmental) which have arisen as a result of society's technical mastery of the natural environment. It has also looked at the ways in which the natural environment has been transformed into 'humanised environments' such as the urban and built environments. At the same time, it explored the implications of how such 'natural' environments as the 'countryside' are a result of present and past collective human transformation. The 'environment' for twentieth-century social theory is thus not confined to the 'natural' environment. Nor is 'nature' confined to 'nonhuman nature', for, as thinkers from Freud to the Frankfurt School have suggested, one cannot talk of the latter without also talking about 'human nature'. Finally, twentieth-century social theory has also raised questions as to the status, value and direction of modernity and the Enlightenment. The price of modern progress, namely the 'disenchantment of nature' and an almost exclusive instrumental cultural and economic valuation of the natural environment, is increasingly regarded as being in need of re-examination in discussions about what Habermas has called 'the unfinished project of modernity'. Thus the discussion of the environment, our relations to it, its meanings and status, are and have been a central part of debates about and within modernity and the legacy of the Enlightenment (see Hayward, 1995).

What marks both Habermas and Giddens and to a lesser extent the Frankfurt School is their stress on the structural, institutional causes of environmental problems. While there is some slippage and ambiguity within Giddens's work, as Goldblatt (1996) argues, between citing industrialism or capitalism as the cause of these problems, Habermas at least is clear in identifying the economic and political structures of the globalising capitalist world order as the ultimate source of the ecologically unsustainable character of modern (and modernising) societies.

## Summary points

- In social theory from the nineteenth to the twentieth century, an appeal to the 'naturalness' of principles or view of social relations has often been seen as indicating that such relations are 'given' that is, cannot be changed by human will.

- Classical sociology had relatively little to say about the natural environment, beyond seeing 'environment' as the opposite of 'culture', which was the main focus of social theory.

- That one cannot discuss nonhuman nature without reference to human nature has been a concern of twentieth-century social theory from Freud to the Frankfurt School.

- Critical theory's assessment of the Enlightenment and modern, industrial societies was the first attempt to systematically analyse the natural environment and its relation to human social practices, as part of its critique of the modern social order. The basic view of critical theory is that the domination and exploitation of the natural environment leads to the domination and exploitation of humans.

- Habermas's social theory holds that an instrumental valuation and relationship to the natural world is unavoidable, but he does see the rise of green politics and a concern for environmental protection and preservation as positive developments.

- The work of Anthony Giddens focuses on the urban experience of modernity as central in explaining the rise of environmental concern, which he says can be linked to a defence of tradition and particular forms of collective and individual identity. Threats to the natural environment are modern risks which arise particularly as a result of globalisation.

- For recent social theorists, Habermas and Giddens in particular, while their theories do shed light on the interaction between human societies and their nonhuman environments, there is an unresolved question regarding the proper place of 'wilderness' areas or of natural processes such as global hydrological cycles, nitrogen cycles, ozone production, within an environmentally aware social theory.

# Further reading

## The Frankfurt School/critical theory

For an overview of the overlaps and tensions between Critical Theory and environmental issues see Stephen Vogel, *Against Nature : The Concept of Nature in Critical Theory* (New York: State University of New York Press, 1996); Andy Dobson, 'Critical Theory and Green Politics' in Andrew Dobson and Paul Lucardie (eds), *The Politics of Nature* (London: Routledge, 1993); and Matthew Gandy, 'Ecology, Modernity, and the Intellectual Legacy of the Frankfurt School', *Philosophy and Geography*, 1: 1, 1996.

## Habermas

Habermas has directly discussed the place of the nonhuman world in his (1982) 'A Reply to My Critics', in J. Thompson and D. Held (eds), *Habermas: Critical Debates* (London: Macmillan), and his assessment of ecological politics in his (1981) article, 'New Social Movements', *Telos*, 49. For an early assessment of Habermas's social theory and nature see, J. Whitebook, (1979) 'The Problem of Nature in Habermas', *Telos*, 40: 41–69, reprinted with a new introduction in D. Macauley (ed.), *Minding Nature: The Philosophers of Ecology* (New York and London: Guildford Press, 1966).

Other commentaries and critical assessments of Habermas include: S. Vogel, 'Habermas and the Ethics of Nature', in R. Gottlieb (ed.), *The Ecological Community: Environmental Challenges for Philosophy, Politics and Morality* (London: Routledge, 1997); C.F. Alford, *Science and the Revenge of Nature* (Tampa: University of Florida Press, 1985); and P. Dickens, *Society and Nature: Towards a Green Social Theory* (Hemel Hempstead: Harvester Wheatsheaf, 1992).

## Giddens

While one can find references to nature and the environment throughout Giddens's work, his most sustained treatment of the topic can be found in the following books: *The Consequences of Modernity* (Cambridge: Polity, 1990) *Modernity and Self-Identity* (Cambridge: Polity, 1991); and *Beyond Left and Right* (Cambridge: Polity, 1994).

David Goldblatt's. *Social Theory and the Environment* (Cambridge: Polity, 1996) devotes two chapters to Giddens (chs 1 and 2) and one to Habermas (ch. 4).

# Gender, the nonhuman world and social thought

- Gendered hierarchies in Western thought and culture
- Ecofeminist spirituality/essentialist ecofeminism
- Materialist ecofeminism
- Ecofeminist political eonomy
- Resistance Ecofeminism

## Introduction

Up until the modern era, the idea of the inequality between men and women and the subservience of women to men as 'natural', something 'given' and beyond human powers to alter was a taken-for-granted perspective. Women were held to be physically (and psychologically) weaker than men and seen as occupying a position somewhere below 'man' but above 'animals' or nature. It is these (and other) historical and conceptual connections between women and nature that makes the adoption of a gendered approach to the discussion of social theory and the environment not just interesting but absolutely essential. As will become clear in later discussion of ecofeminist social theory, social theorising about the environment is not a gender-free zone.

The connection between gender and the environment within social theory is something that has received much attention in the last three decades or so. Tracing this relationship has been a key, if not always a central part, of the feminist movement and feminist social theory since the 1960s. However, the connection between gender, environment and social theory has its origins in the late eighteenth century with Mary Wollstonecraft's book *A Vindication of the Rights of Women* (published in 1792) and reactions to it. In this seminal book, Wollstonecraft argued for the extension of some limited equal rights to (some) women, such as the right to hold property, capital and education. However, her proposal for 'women's rights' provoked a reaction which starkly reveals the dominant view of women at the time. This was the publication of a book in response to Wollstonecraft, entitled *A Vindication of the Rights of Brutes*. The basic position of the response was that if women were to have rights, then why should animals not also have them? Here, an explicit connection was made between the status of women and animals, a similarity between women and animals which was to last both historically and conceptually up until recent times and the creation of the feminist movement which challenged such 'sexist' assumptions (although as we shall see this is not the whole story).

It may seem odd to attempt to link ecology and feminism, since as Mellor points out, 'While feminism has historically sought to explain and overcome women's association with the natural, ecology is attempting to re-embed humanity in its natural framework' (1997: 180). However, for Salleh (1997) this identification of women with nature, upon which the whole edifice of hierarchical dualisms within Western culture and thought has been built, should be welcomed by feminists. As she puts it, 'Feminists should not fear the double-edged metaphor of Mother=Nature. This nexus both describes the source of women's power and integrity, and at the same time exposes the complex of pathological practices known as capitalist patriarchy' (1997: 175). It is to these gendered dualisms that we turn next.

## Gendered hierarchies in Western thought and culture

Following ecofeminist social theorists such as Plumwood (1993) and Merchant (1990), we can begin to understand the relationship between gender and the environment by firstly noting a series of gendered dualisms that exist within, and define, Western culture as a patriarchal culture. The origins and effects of patriarchy are central to any feminist analysis, and it is worthwhile to note Marylin French's definition of it as 'an ideology founded on the assumption that man is *distinct from the animal and superior to it*' (quoted in Zimmerman, 1987: 25; emphasis added). According to this line of analysis, Western patriarchal culture is based on a gendered separation of 'culture' from 'nature' such that male attributes and values are associated with culture, while female attributes and values are associated with nature. However, it is not just that there is this dualism within Western culture, it is also the case that both historically and conceptually male attributes have be seen as not just separate from but also 'superior' to female ones. For some ecofeminists, such as Plumwood (1993) and Merchant (1990), the creation and maintenance of these sets of hierarchical gendered dualisms has its roots in Judeo-Christianity in general, and the 'domination of nature' thesis in particular, discussed earlier in Chapter 2.

The association of women and nature has historically produced the following sets of hierarchies (dualisms or binary oppositions), such that items on the left hand side are accorded more importance or value than those on the right.

*Gendered Hierarchies in Western Thought and Culture*

Culture / Nature
Men / Women
Human / Non-human
Reason / Emotion
Mind / Body
Abstract / Concrete
Objective / Subjective
Public / Private
Production / Reproduction
Rationality / Intuition
Competition / Co-operation
Violence / Non-violence

'Human' or the 'really' human came to be associated and identified with those 'male' characteristics and properties on the left-hand side. 'Female' characteristics on the right-hand side have been historically viewed as not representing what is 'truly' or distinctly human about human beings on account of being too closely tied to nature, the body, animality, sensuality, emotions, etc. This is the basic proposition and historical analysis of most feminist critiques of sexism and patriarchy. Namely that Western culture (and many non-Western ones too) privileges certain (i.e. 'male') attributes and properties (reason, abstract thought, mind, culture, production) above others (i.e. 'female') attributes and properties (emotion, concrete thought, the body, nature, reproduction). As Merchant points out, 'Anthropologists have pointed out that nature and women are both perceived to be on a lower level than culture, which has been associated symbolically and historically with men. Because women's physiological functions of reproduction, nurture, and child rearing are viewed as closer to nature, their social role is lower on the cultural scale than that of the male' (1990: 143).

The implications of this are substantial for the examination of the relationship between social theory, social practice and the environment. Since this set of dichotomies/dualisms is at the heart of Western culture, this means that one cannot examine 'nature' or the environment in Western thought without adopting a gender perspective. That is, one needs to look at gender in examining environmental issues generally, since social-environmental interaction, views of the environment and its significance for society, are not 'gender-free' zones. For example, the very language used in social theory and in everyday discourse about the environment, ecological relations and the interaction between society and the nonhuman world is saturated with gendered terms. For example we speak of 'virgin lands', the 'rape of the wild' (Collard, 1988), the 'despoliation of nature', 'Mother Earth' etc., all of which are clearly gendered terms. Thus, some feminist social theorists have made the gendered (i.e. socially constructed) connection between women and nature explicit, and in so doing began the process of developing an *ecofeminist* perspective. As Karen Warren has noted,

> As I see it the term ecofeminism is a position based on the following claims: (i) there are important connections between the oppression of women and the oppression of nature, (ii) understanding the nature of these connections is necessary to any adequate understanding of the oppression of women and the oppression of nature; (iii) feminist theory and practice must include an ecological perspective; and (iv) solutions to ecological problems must include a feminist perspective.
> (Warren, 1987: 4–5)

The joining together of the ecological and feminist movement is for some ecofeminists a condition for success of both struggles. That is, feminists cannot achieve their ends within incorporating ecological concerns, and ecological aims will be frustrated without a feminist dimension. This strong argument for connecting feminism and ecology has been made by Rosemary Reuther. According to her, 'Women must see that there can be no liberation for them and no solution to the ecological crisis within a society whose fundamental model of relationships continues to be one of domination. They must unite the demands of the women's movement with those of the ecological movement to envision a radical reshaping of the basic socio-economic relations and the underlying values of this society' (quoted in Pietilä, 1990: 200–1).

Perhaps the best way of illustrating the complexities, variety and potential insights to be gained from paying close attention to gender and the environment within social theory is by outlining some of the main lines of thinking within ecofeminism. There are at least three main schools of thought which adopt this explicitly gendered approach to theorising the environment and social-environmental relations. These different approaches to ecofeminist social theory are discussed below.

## Ecofeminist spirituality/essentialist ecofeminism

The basic argument of this school of ecofeminist thought is that a necessary condition for solving the ecological crisis and attaining more equality for women is to reverse the gendered dualisms indicated above. There is a strong sense within this branch of ecofeminism which states that the ecological crisis is not simply because of the anthropocentrism (i.e. human-centredness) of the 'modern' or Enlightenment worldview, but can be traced to the androcentrism (i.e. male-centredness) underpinning it. As Plant explains,

> the world is rapidly being penetrated, consumed and destroyed by this
> man's world – spreading across the face of the earth, teasing and tempting
> the last remnants of loving peoples with its modern glass beads –
> televisions and tanks . . . As the Amazon rainforest is bulldozed to
> provide cheap beef for American hamburgers, the habitat of peoples who
> once lived and loved with this earth and the fragile womb of planet Earth
> is dealt yet another killing blow. This 'man's world' is on the very edge
> of collapse.
>
> (Plant, 1989: 1–2)

And if it is this 'man's world' which is the problem, then the solution is the creation of a 'women's world', since women are 'closer to nature' and thus better suited than men as guardians and protectors of the natural environment. This is the 'essentialist' element: *women are essentially closer to nature because of their particular biological natures and abilities/capacities.* Women like nature reproduce, are life-givers and nurturers, they are in touch with 'natural cycles' such as menstruation, which have affinities with natural cycles, such as those of the seasons. It is important to note here that what is being criticised (and this is true of all schools of ecofeminist thought) is not 'men' per se as individuals or as a group. Rather the main point of criticism is 'male' forms of thinking, institutions and practices which have led both to the degradation of the natural world, and the oppression of women and the denigration of female values and attributes. Again as Plant puts it,

> Making the connection between feminism and ecology enables us to step
> outside of the dualistic, separated world into which we were all born.
> From this vantage point, this new perspective, we begin to see how our
> relations with each other are reflected in our relations with the natural
> world. The rape of the earth, in all its many guises, becomes a metaphor
> for the rape of woman, in all its many guises. In layer after layer, a truly
> sick society is revealed, a society of alienated relationships all linked to a
> rationalization that separates 'man' from nature.
>
> (Plant, 1989: 5)

A central part of this school of ecofeminism is the practical importance
of 'earth-bonding' spirituality and rituals, and the need for men to
discover or become more in tune with their 'female side', that is to
become more associated with what have been traditionally regarded as
'female' values and practices such as emotion, caring, intuition (the
items on the right hand side of the list above). Some of these rituals take
the form of a return to 'Wicca' or 'witchcraft' (which is not the same as
Satanism), as part of a return to older more woman-centred, natural or
pagan religions which Judeo-Christianity had destroyed. In opposition to
the Judeo-Christian God which is male and located in heaven (i.e. not on
Earth) and directed towards humanity, ecofeminist spirituality sees
divinity, meaning and spirituality as existing throughout 'Mother Nature'
and not as the special preserve of humans alone.

The main points of ecofeminist spirituality can be summarised as
follows:

1  There needs to be a reversal of the binary oppositions within Western
   culture, such that 'female' attributes, concepts and ways of thinking
   and acting form the basis of a new ecologically sensitive, life-
   affirming culture.
2  Action and real experience of the natural world is prioritised over
   abstract theorising about it.
3  Women rather than men ought to represent nature.
4  Personal (inner) change must precede political or social (external)
   change.

However, like any subdivision of a larger body of social theory, this
branch of ecofeminism has been criticised by other schools of
ecofeminism, particularly materialist ecofeminism (as well as by other
schools of feminist thinking more generally). One of the main criticisms
levelled concerns the 'essentialist' view of women. Here, it has been
claimed that there is a confusion of sexual characteristics (biology) with

social roles (gender). 'Biology is not destiny' as the feminist slogan has it. Thus ecofeminist spirituality is found to be offering a biological determinist view of women, which is dangerous and misleading. For example, one can ask about the status of women who do not fulfil their 'biological duty/function' in having children. According to the logic of essentialism, women are seen as essentially mothers, which seems to imply that women who choose not be mothers are somehow denying their 'essential nature'. The latter is a conservative or traditional view of women which other feminists are struggling to deconstruct. In short, it seems as if essentialist ecofeminism simply wants to 'celebrate' the very attributes and views of women that have (a) been formed under sexist conditions and (b) been traditionally used to keep women down.

A second problem following on from the first is the confusion between the *feminine* as opposed to the *feminist* character of this form of ecofeminist social theorising. In urging the celebration of qualities such as mothering, caring, feeling, and nurturing, by basing it on an appeal to the essential, biological, psychological 'nature' of women, they advance a revaluation of the 'feminine' as opposed to a feminist politics aiming for equality between women and men. An illustration of this is in the way green politics is often seen to be less sexist and more 'pro-women' than other political ideologies such as liberalism or socialism. However, as Mary Mellor points out, 'Where male green thinkers claim that a commitment to feminism is at the heart of their politics, this often slides into a discussion of feminine values' (Mellor, 1992b: 245). Celebrating the feminine (as if this were a biologically 'given' concept) according to Mellor demonstrates a lack of awareness of the fundamental distinction between biology and gender. As she puts it, 'The feminine is not the missing half of the masculine; the feminine is what men need to create the masculine in patriarchal culture' (Mellor, 1992a: 81). A social theory and politics based on the feminine, as opposed to a feminist perspective, does not challenge sexism and gender inequalities, rather it serves to reinforce them. Thus essentialist ecofeminists may succeed in 'saving the planet' (and men) at the cost of female liberation and equality.

A third and final problem lies with a whole series of questions relating to defining 'female' characteristics. How can we determine what are 'female' traits if women's lives have been determined, structured and influenced by patriarchy and sexism? Are submissiveness and unassertiveness 'female' characteristics and therefore something to be celebrated and encouraged? As Mellor puts it, 'Feminists have long argued that until women have control over their own fertility, sexuality and economic circumstances, we will never know what women "really" want or are' (1992b: 237). Taking up this challenge has been one of the main aims of materialist ecofeminism which we turn to next.

## Materialist ecofeminism

In opposition to many of the tenets of essentialist ecofeminism stands what I have termed 'materialist ecofeminism', which represents a more nuanced and sophisticated approach to the relationship between social theory, gender and the environment. For Mary Mellor (1992b: 162), one of the main social theorists in this area, 'despite the influence of cultural and spiritual feminism, ecofeminism is necessarily a materialist theory because of its stress on the immanence (embodiedness and embeddedness) of human existence'. Ariel Sallah echoes this point, noting that,

> It is nonsense to assume that women are any closer to nature than men. The point is that women's reproductive labour and such patriarchally assigned work roles as cooking and cleaning bridge men and nature in a very obvious way, and one that is denigrated by patriarchal culture. Mining or engineering work similarly is a transaction with nature. The difference is that this work comes to be mediated by a language of domination that ideologically reinforces masculine identity as powerful, aggressive, and separate over and above nature.
>
> (Salleh, 1992: 208–9)

Here the connection between women and nature is not on the basis of some 'essentialist' or 'biological' grounds; rather it is the fact that women's work (including reproductive work) means that they are closer to nature. Whereas the essentialism of ecofeminist spirituality located the connection between women and nature in *sex* (biological characteristics of women), materialist ecofeminism locates the connection in *gender* (social constructions of practices, characteristics and roles based on sex). Women and nature both suffer at the hands of patriarchy and industrial capitalism. That is, *what unites women and nature, is not the biological closeness of women to nature, but the fact that both are exploited and oppressed by male, sexist culture, its institutions, values and practices.*

Thus the root of the connection (both historical and conceptual) between women and nature lies in their material exploitation, as a result of the sexist organisation of society and the economy. As Mellor puts it, 'A feminist green politics must begin with women's work of nurturing and caring and the sexual division of labour that largely excludes men from that work' (1992b: 278). In material terms women are assigned to the (disvalued or non-valued) sphere of 'reproduction' (child-rearing, food preparation, home-making, nurturing, etc.), while men are assigned the (valued) sphere of production (producing commodities for the market, earning income). The dis-valued or non-valued status of 'women's reproductive work' can be easily seen in that the work women do in the home – food preparation, cleaning, child-rearing, caring, comforting, etc. – is not paid. From the perspective of orthodox theories of economics, which are themselves gendered (surveyed in the next chapter), women's reproductive and nurturing work is 'free' (just as environmental services are also seen as 'free').

The 'life-affirming' character of reproductive work is used by Salleh to explain why in Western social theory and history this gendered sphere of activity, and the characteristics and values associated with it, have been downgraded. As she puts it, 'In the Eurocentric tradition, not "giving life" but "risking life" is the event that raises Man above the animal. In reality, reproductive labour is traumatic and highly dangerous . . . birthing . . . is an experience that carves the meaning and value of life into flesh itself' (1997: 39). Hence life-taking and its associated activities and values of violence and war are, within Western culture, seen as an essential aspect of what is distinctively 'human' as opposed to 'nonhuman'. This also partly explains the association of men and manliness (and thus 'humanness') with violence and warfare, something Freud sought to explain as discussed in the last chapter. In this way 'human' becomes associated with 'maleness'. At the same time life-giving, birth, reproduction and nurturing are not seen as something that is distinctively 'human', since these biological or 'natural' activities are something we share with the nonhuman world.

Materialist ecofeminism is, unlike ecofeminist spiritualism, orientated towards reconfiguring the material basis of human society (covering the formal and informal economy, the nature of work, reproductive relations, and the material exchange between the 'total human economy' and the natural environment). Figure 5.1 below outlines the materialist ecofeminist position.

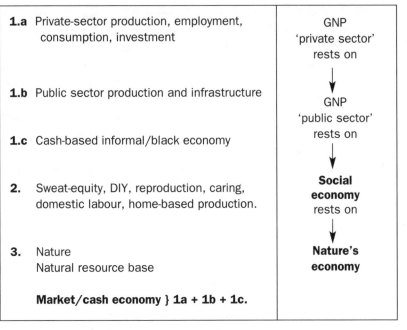

**Figure 5.1** *The Materialist Basis of Human Society*

*Source:* Adapted from Henderson, H. *et al.* (1986) 'Indicators of no Real Meaning', in Ekins, P. (ed.) (1986) *The Living Economy: A New Economics in the Making*, London: Greenpoint, p.33.

## Ecofeminist political economy

The full force of the materialist ecofeminist position demands nothing short of the radical transformation not only of the economy, but also a radically new theory of economics. While the next chapter will outline some of the shortcomings of orthodox theories of economics from a 'green' perspective (which includes some of the insights of materialist ecofeminism), some of the criticisms levelled at these theories can be listed. The changes materialist ecofeminism requires within economic theory include: the creation of meaningful indicators of human 'well-being' rather than abstract measures of 'economic growth'; the reconceptualisation of central economic categories such as 'work' (to include reproductive work); transcending the 'public/private' division by extending the notion of the 'economy' to include the informal, social and domestic economy; and recognising the dependence of the 'total human economy' on the natural environment. The basic ecofeminist political economy position is as follows:

*Ecofeminist political economy*

1. Sphere of Production (industry, formal economy)

rests on

2. Sphere of Reproduction (nurturing, informal economy)

rests on

3. Nature's economy (natural resources)

The idea of dependence is central to the ecofeminist materialist position in their critique of orthodox, gender-blind economic theory and practice (and 'malestream'/mainstream political theory and practice). Dependence, vulnerability and the inherent neediness of humans are central ideas and realities of the 'human condition' which have been ignored and/or denied within (male) economic, political and social theory and practice. Because standard or conventional economic theory and practice does not take into account the double dependence of humans (our dependence on each other and on the natural world), the individual and collective vulnerability of human beings is denied.

The result of this is that gender-insensitive and environmentally-blind economic theory and practice have resulted in a situation whereby the biological and ecological character of human collective and individual life are simply unacknowledged. As Mellor puts it, 'As a result of women's private and unacknowledged labour we have a *public world constructed on the false promise of an independently functioning individual, with the nurturing, caring and supportive world hidden, unpaid and unacknowledged*' (1992b: 239–40). Thus we have an economy in which childcare considerations are simply not seen as 'appropriate' or central when making economic decisions. These are relegated to the 'private' and supposedly non-economic realm of the home. At the same time the biological and psychological needs of humans are likewise not given the prominence they deserve. An example of this is the argument that the '24-hour working day' as an economic ideal is completely out of synch with the biological and psychological needs of human beings, and is not only impossible and undesirable, but based on a completely false picture of human beings which does not acknowledge their inherent neediness and vulnerability. Finally, orthodox economics has largely ignored the contribution of the natural environment to the human economy and its dependence on that environment.

Ecofeminist materialist political economy stresses the experience and labour associated with *reproduction*, the private, unvalued but fundamental life-sustaining work women perform. This life-sustaining focus is particularly evident in Salleh's view when she makes clear that, 'the embodied materialism of ecofeminism is a "womanist" rather than a feminist politics. It theorises an intuitive historical choice of re/sisters around the world *to put life before freedom* . . . Ecofeminism is more than an identity politics, it reaches for an earth democracy, across cultures and species' (1997: ix–x; emphasis added). This adoption of a 'womanist' rather than a 'feminist' stance is motivated by a desire to make connections with women in the Southern, developing world, whose concerns, problems and issues are not articulated by the privileged, urban, affluence-based discourses of Northern/Western feminism.

Salleh offers powerful criticisms of Northern, liberal feminism from a materialist ecofeminist standpoint. She criticises Northern, affluent feminist concerns with individual self-realisation, its Eurocentrism and insensitivity to Southern women's concerns, its anti-reproductive bias, and ultimate blindness to its position within global capitalism. Salleh suggests that 'For too many equality feminists, the link between their own emancipated urban affluence and unequal appropriation of global resources goes unexamined . . . Much of the energy that went into abortion campaigning was clearly a sublimation of this hostility toward the problematic mother. The unreality of mothering experiences to many feminists did not help theorisation . . . The hope is that feminism's ideological immaturity will be remedied as this generation of career women take up mothering themselves, and draw that learning into feminist thought' (1997: 104). She sees Northern liberal/equality feminism as the product of what Marcuse called the 'repressive tolerance' of patriarchal capitalist states, in which feminist issues are 'co-opted' and thus neutralised, and feminist activists become 'femo-crats'. Thus, materialist ecofeminism is suggested as a maturing or development of feminism both as a form of social theory and a political movement.

## Resistance ecofeminism

While not a 'school' in the sense that essentialist and materialist ecofeminism are, there is a third strand of ecofeminist thought. Overlapping with some of the concerns of the other two, though more on the materialist than essentialist side, it does represent a distinctive 'voice'

# The Economic Totem Pole

**What you see is definitely *not* what you get if you look at modern industrial economies in the traditional way. Underlying the visible monetary economy is a whole *non*-monetary area of activity which is both invisible and undervalued. Without the strong shoulders of the bottom characters in this totem pole the market economy at the top would quickly tumble down.**

## Official market economy

All transactions and relationships involve money – including wages, consumption, production, investment and savings.

## Government expenditure

Money spent by government on social security, defence, education and infrastructure like roads, bridges, airports, sewers and public transport.

## Underground economy

Cash transactions which are hidden to avoid taxes or illegal like drug trafficking, prostitution and pornography.

## Social economy

All non-market economic activities. Includes subsistence farming, housework, parenting, volunteer labour, home healthcare and DIY. Also includes barter or skill exchanges. In Northern economies the informal economy is estimated to be one-and-a-half times the size of the visible market economy.

## Mother Nature

The natural resource base is the largest and most basic support for the monetary economy. All economic activity depends on the survival of healthy, natural ecosystems.

**Figure 5.2 *'The Economic Totem Pole'***

*Source:* Polyp, P.J., in *New Internationalist*, April, 1996

---

## Box 5.1

### Summary of materialist ecofeminism

1 Stresses the importance of spheres of production and reproduction, and critical of existing, gender-blind political and economic theories, institutions and structures.

2 Strongly 'feminist' as opposed to 'feminine'.

3 Emphasises the dependence of humans on nature, and also the real work women are socially associated with and which is vital to human society, i.e. reproductive work, caring, nurturing, home-making.

4 The ecological restructuring of the economy requires reconstructing the relationship between the spheres of 'production' (public) and 'reproduction' (private), so that the latter takes precedence.

5 This ecological restructuring requires a new materialist ecofeminist theory of economics, in which central categories of economic thought need to be reconceptualised.

6 Stresses the biological (and psychological) neediness, vulnerability and dependency of humans.

---

and perspective within ecofeminism. This final stream of ecofeminism is characterised by its practical political concerns and while it does have relevance to the developed world, its origins and main focus lie in the 'developing' world.

A key starting point for resistance ecofeminism is the recognition that women are more concerned about the environment than men, and that women are at the forefront of many environmental struggles. Examples of the latter include the Chipko movement in India, a movement of local women in Uttar Pradesh protesting against commercial logging which was leading to rapid deforestation (Ekins, 1992: 143), and the British Greenham Women's Peace Camp which was a 1980s anti-nuclear movement to remove American nuclear missiles from Britain, and which had strong ecological and feminist aims. From local community movements against toxic dumping, protests against increased road-traffic, resistance to timber-logging, dam-building and other 'mega-developments' in the developing world, women are either in leading positions or make up the bulk of support for these various environmental resistance movements.

For certain environmental struggles, particularly when they have to do with health issues, or subsistence livelihoods, it is often the case that women are at the forefront. Also, in different environmental struggles or environmental issues, women may have a greater vested interest than men (on the issue of population 'control', for example). In other words, there are aspects of environmental protection which are gendered in a way other areas of environmentalism such as biodiversity loss and global climate change are not. Where environmental problems affect human health such as toxic-dumping, electro-magnetic radiation and anti-smog campaigns, women are commonly at the forefront, particularly when it is children's health that is at risk. What is interesting about women's involvement in environmental campaigns is that often it becomes a training-ground for more mainstream 'feminist' demands for greater equality, access to employment, wages and general standards of welfare and political, economic and social respect. As Martin-Brown observers, 'environmentalism has become the "Trojan horse" for the engagement of women in the political process' (Martin-Brown, 1992: 707).

Struggles and movements against genetic engineering and biotechnology are another focus for environmental resistance in which women have, according to Vandana Shiva a particular interest. According to her, 'Capital now has to search for new colonies to invade, exploit, and spoil for further accumulation. These new colonies are the interior spaces of the bodies of women, plants and animals . . . Biotechnology as the handmaiden of capital in the post-industrial era creates the possibility to colonise and control that which is free and self-regenerative' (1992: 13).

It is generally regarded that only by giving women reproductive rights will major environmental and developmental problems be averted in the developing world. Here the traditional feminist demand for women to be given control over their fertility, and environmentalist claims of the positive relationship between population growth and environmental degradation dovetail into one another. It is perhaps on this issue if no other that 'ecofeminism' can be considered as a synthesis of feminism and environmentalism. Studies have demonstrated that there is a causal connection between women's equality in general (and reproductive rights in particular) and population control and environmental protection; equality for women means having control over their own reproduction.

Addressing the claims of Malthus, discussed earlier, materialist ecofeminism rejects the idea that population growth is the sole cause of global environmental problems. Many ecofeminists, like Bandarage, see contemporary 'neo-Malthusianism' as the dominant ideological analysis and approach to the global ecological crisis. She suggests that, 'Like Malthus, contemporary Malthusian analysts who work within the population control paradigm advocate population stablization as a substitute for social justice and political-economic transformation' (1997: 6). However, as she puts it, 'growing global economic inequality, not population growth, is the main issue of our time' (1997: 12).

Bandarage also uncovers the violence underpinning population control rhetoric and practice, noting how 'military metaphors that "declare war", "target" and "attack" "over population" with an "arsenal" of new drugs have become the standard language of global population control' (1997: 65). This gendered rhetoric and way of thinking about population issues is particularly striking in that its 'masculine', not to say 'macho', character is quite obvious. This violent and warlike mode of thinking and acting leads, according to Bandarage to the situation where, 'Aggression and conquest rather than compassion and care drive the population control establishment and the larger model of technological-capitalist development that it represents' (1997: 103).

In many so called 'Third World' countries, it is women who have to provide, tend and prepare food. With modernisation, imported technologies, increasing immersion into the global market, etc., it is women who lose out more, as they are displaced from subsistence labouring by mechanisation and pesticides. Women in Third World economies are increasingly forced to produce food on marginalised land as previously farm land is intensively cultivated for the production of cash crops for export. This not only increases the burden on them but degrades the surrounding land. Women as the primary 'land managers' and workers in agriculture in the South mean that any attempt to implement sustainable policies needs to take this fact into account. That is, policies aimed at **sustainable development** or environmental protection must be formulated with women and women's needs in mind. According to Martin-Brown, 'Traditionally, that responsibility [of ecosystem management] has fallen to women. Throughout time and around the world, the traditional role of woman has been to *manage* prescribed resources . . . The complementary historical role of men has been to *enlarge* the available resource base' (1992: 707). From this she takes the view that sustainable development will depend on how women can be empowered to manage ecosystems.

At the same time, in the developing world there is a connection between poverty and environmental degradation. Yet while this is a generally recognised relationship (Doyle and McEachern, 1998: 77; Goldblatt, 1996), there is also a gender aspect to this relationship which needs equally to be recognised. The fact is that women (and children) are the 'poorest of the poor' in the developing world, and thus suffer more than anyone else from the effects of poverty, poor environments and environmental degradation. Thus within the context of sustainable development in the South, joining development with environmental protection, one needs to be aware of the gendered distribution of environmental and economic burdens/costs, which mean that it is women as a group which suffer the most and therefore whose needs are greatest. As Martin-Brown notes, 'Women and the environment are the "shadow subsidies" which support all societies' (1992: 717).

According to Vandana Shiva (1988) women are in the vanguard against Western forms of **modernisation** and its damaging ecological and socio-economic effects in the South. For her, women in the developing world offer resistance to what she calls the 'colonisation' of the developing world, by the imposition of a Western and male view of progress, economic vision, institutions and modes of thinking and acting about the natural environment. As Doyle and McEachern note in discussing Shiva's argument, women for her in 'subjugated cultures have been direct activists in opposing modernisation in parts of the third world' (1998: 51). At a more conceptual level, Shiva advances an ecofeminist critique of the Enlightenment (the original source as it were for modernisation theory and practice). For her:

> Throughout the world, a new questioning is growing, rooted in the experience of those for whom the spread of what was called 'Enlightenment' has been the spread of darkness, of the extinction of life and life-enhancing processes. A new awareness is growing that is questioning the sanctity of science and development and revealing that these are not universal categories of progress, but the special projects of modern western patriarchy . . . *The violence to nature, which seems intrinsic to the dominant development model, is also associated with violence to women* who depend on nature for drawing sustenance for themselves, their families, their societies. This violence against nature and women is built into the very mode of perceiving both, and forms the basis of the current development paradigm.
>
> (Shiva, 1988: xiv; emphasis added)

For Shiva, then, resistance to large-scale projects of modernisation, such as the Narmada Dam in India (Ekins, 1992: 114), or protests against the patenting of seeds and genetic information by Western biotechnology industries (Purdue, 1995), can also be seen as part of a wider and deeper resistance to a particular Enlightenment or 'modern' (i.e. Western) ways of thinking, valuing and acting.

Thus this 'resistance' ecofeminism shares some of the concerns of materialist ecofeminism, but its critique of ecological degradation and alternatives to the status quo are premised on practical experience of ecological struggles. In its developing-world forms resistance ecofeminism constitutes a rejection of Western forms and models of modernisation.

## Conclusion

Feminism has made the relationship between human society and the natural environment central to its concerns, more so than any other twentieth-century social theory. Eco-feminism, as a subbranch of feminism concerned with ecological issues, highlights the role of gender in social-environmental relations. Different strands of ecofeminism, such as essentialist/spiritual, materialist and resistance each offer their own critical analysis of the connection between the oppression of women and the degradation of the natural environment, and their own (sometimes competing) alternatives to what they regard as contemporary anti-women and anti-environmental political, social and economic arrangements.

## Summary points

- Exploring the relationship between gender and the environment has been the central contribution of feminist social theory to the study of social-environmental issues. Social theorising about the environment is not a 'gender-free' zone.

- Western culture is based on a set of gendered dualisms such that (a) certain values, principles, characteristics and activities are either 'male' or 'female' and (b) those that are 'male' are regarded as both separate and superior to those associated with 'female'.

- Essentialist or spiritual ecofeminism is based on the claim that women are 'naturally' closer to nature than men, and that if the cause of ecological problems is men and male culture, then the solution is the creation of a women-centred society.

- Essentialist ecofeminism has been criticised for confusing biological sex (which is 'given') with gender (which is socially constructed), and in being more 'feminine' than 'feminist'.

- Materialist ecofeminism begins from the observation that what connects women and nature is that both are exploited within patriarchal or male-dominated society.

- It stresses the biological embodiedness and ecological embeddedness of human beings, and draws attention to the vulnerability, neediness and dependence of humans.

- It highlights the role of women's unpaid and undervalued reproductive labour in meeting human needs and the ways in which 'malestream' social, political and especially economic theory and practice, ignore both this vulnerability and women's work.

- Ecofeminist political economy calls for the radical reconceptualisation of 'malestream' or orthodox economic theory especially central economic terms, while in practice calling for the radical restructuring of the economy.

- Resistance ecofeminism links the feminist and ecological movement in terms of common political aims, such as defending women's reproductive rights and protecting women from poverty and degraded environments.

- Resistance ecofeminism also builds on the observation that certain environmental issues (population, health-related, children-related and the link between poverty and environmental degradation) seem to be more gendered in terms of actual social support by more women than men, than others.

## Further reading

For a general overview of ecofeminism, see Val Plumwood's *Feminism and the Mastery of Nature* (London: Routledge, 1993), Karen Warren (ed.) *Eco-Feminsm* (London: Routledge, 1995), and A. Collard, *Rape of the Wild: Man's Violence against Animals and the Earth* (Indianapolis: Indiana University Press, 1988). For shorter overviews see ch. 2 of Timothy Doyle and Doug McEachern's *Environment and Politics* (London: Routledge, 1998), ch. 5 of Andy Dobson's, *Green Political Thought* (London: Routledge, 2nd edn, 1995), Judy Evans, 'Ecofeminism and the Politics of the Gendered Self', in A. Dobson and P. Lucardie (eds), *The Politics of Nature* (London: Routledge, 1993).

On 'essentialist' ecofeminism, see the collection of writings edited by Judith Plant, *Healing the Wounds: The Promise of Eco-Feminism*, (Philadelphia: New Society Publishers, 1989).

For a brief overview of materialist ecofeminism, see my 'Feminism and Socialism (and Ecology) Back Together Again? The Emergence of Ecofeminist

Political Economy', *Environmental Politics*, 7: 4 (1998). Fuller elaborations of materialist ecofeminism can be found in: Mary Mellor, *Breaking the Boundaries: Towards a Feminist, Green Socialism* (London: Virago, 1992), and her latest book, *Feminism and Ecology* (Cambridge: Polity Press, 1997); Ariel Salleh, *Ecofeminism as Politics: Nature, Marx and the Postmodern* (London: Zed Books, 1997) and her 'Nature, Woman, Labor, Capital: Living the Deepest Contradiction', in M. O'Connor (ed.), *Is Capitalism Sustainable?: Political Economy and the Politics of Ecology*, (New York and London: Guildford Press, 1995); and Asoka Bandarage, *Women, Population and Global Crisis: A Political-Economic Analysis* (London: Zed Books, 1997).

On 'resistance' ecofeminism, see Vandana Shiva, *Staying Alive: Women, Ecology and Development* (London: Zed Books, 1998), *Monocultures of the Mind* (London and Penang: Zed Books and Third World Network, 1993), and Joan Martin-Brown (1992), 'Women in the Ecological Mainstream', *World Development* XLVII: 4.

# 6 ⬤ The environment and economic thought

## Introduction

This chapter takes up some historical themes discussed in Chapters 2 and 3, as the starting point for an examination of the relationship between economic thought and the environment. The reason for focusing on economic thought is defended on the grounds that, firstly, economic thought *is* a form of social theory. Different forms of economic thought are premised on particular analyses of society and views on alternative social arrangements. Secondly, different forms of economic theory are based on particular moral principles, including views and conceptions of human nature and the value of the nonhuman world. Thirdly, and this will form the bulk of this chapter, of all forms of social theorising, economics has had the most influence on both how the environment and social-environmental interaction has been conceptualised. Fourthly, and following on from the last, economic thought has had (and continues to have) a dominant position within prevailing political and economic institutions which mediate and shape the actual, material relationship between society and environment.

The main aims of this chapter are, firstly, to trace the changing character and role of nature in economic thought; secondly to distinguish political economy from modern discipline of 'economics' and argue that this has important implications for the economic analysis of environmental issues; and thirdly, to look at how 'value' in general and in relation to 'nature' in particular is and has been understood within economic thought.

## The environment and the 'economic problem'

Economics, both as a discipline as well as a particular form of human practice can be understood with reference to what has been called the 'economic problem'. The economic problem refers to the fact that human wants are infinite in comparison to finite or scarce resources (or means) to meet those wants (or ends). Hence economics can be defined as how the economic problem is resolved. At this initial stage it is significant to note that there are two possible ways of resolving the 'economic problem': either we expand resources to meet more and more wants (increase the supply), or we can limit wants in relation to (fixed or limited) resources (decrease the demand). While oversimplifying the issue greatly, it is on this basic distinction that the difference between orthodox economic theory and ecological or green critique of economic theory and practice is based. Thus a basic definition of the concerns of economics can be reduced to two: 'the efficient allocation of available resources and the problem of reconciling available resources with a virtually infinite desire for goods and services' (*Hutchinson Dictionary of Ideas*, 1994: 162).

Central to economics then is the notion of scarcity, the simple fact that resources (natural and human-made) are scarce in relation to infinite wants. Basically, humans can potentially have more wants than can ever be met by the resources available to meet them. Now while resources can include things like 'money', 'labour', 'capital', etc., it is the case that natural resources (that is the useful things humans can get from nature) forms the greatest part of the 'economic problem'. That is, it is the natural environment around us which provides the primary resources upon which the human economy is based. At the same time, the second problem that economics deals with, as indicated above, has to do with the fact that different resources have competing uses. For example, a forest can be logged for the timber it contains, it can be preserved as a 'national park', or it may developed as a housing estate or an airport. Providing methods to enable us to make such choices is central to economic theory and practice.

Given this stress on scarcity, resource limits and so on, it would seem as if economics is close to green or ecological concerns about limits, yet this is not the case. It is also important to note that the etymology of 'ecology' can be defined as the 'economy of nature' that is 'the study of earth's household', while economics derives from the Greek term for 'management of the household'. Yet, throughout the modern history of economics, there has been little real sense of the intrinsic connection between the two, and the fact that economics is embedded in ecology has been something largely ignored.

## Economics as social theory

While it may be seen as odd to include a chapter on economic theory in a book about the environment and social theory, I contend that unless one appreciates it as a form of social theory and practice, one cannot fully understand the relationship between the environment and social theory. The main reason for this is that in the modern period, economic forms of viewing, valuing and understanding the place of the environment within human society have had the greatest practical as well as theoretical effects on social-environmental relations.

Alongside scientific and technological innovation (with which it is closely associated), economic thought and practice (whether it be the 'positive' economics of this century, or the political economy of the last three), has largely created the modern social world, shaped its view of the natural world and focused attention on 'worldly' affairs. Economists, according to Heilbroner can be seen as 'worldly philosophers', since 'they sought to embrace in a scheme of philosophy the most worldly of all of man's activities – his drive for wealth. It is not, perhaps, the most elegant kind of philosophy, but there is no more intriguing or important one' (1967: 14). Its importance as a dominant and dominating form of social theory and practice cannot be underestimated. It is also worth noting how radical economics and the economic worldview was in that the view of society it elaborated, and the particular mechanisms it suggested for holding it together, went against most of the previous history of human societies and received social theory.

One of the central achievements of the 'economic' view of society and social affairs, originating in the seventeenth and eighteenth centuries, was to 'disembed' the economic activities of individuals and groups (production, consumption and distribution) from 'social' or 'political' regulation. That is, modern economic activity and reflection on it was impossible and unnecessary up until the economy emerged as a distinct entity and set of relationships which was separate from religious, political and customary rules, laws, rights and prohibitions. Thus the rise of economic thought must be seen as part of the complex historical processes of change which took place within European societies in the period leading up to, including and continuing beyond the Enlightenment. While not exactly correct, one could say that the birth of economics, the emergence of the 'market economy' governed by prices, voluntary exchange, and not by custom or political-religious rules, can be said to herald the beginning of the modern era.

> The market pattern . . . being related to a peculiar motive of its own, the motive of truck or barter, is capable of creating a specific institution, namely, the market. Ultimately, that is why the control of the economic system by the market is of overwhelming consequence to the whole organisation of society: it means no less than the running of society as an adjunct of the market. Instead of economic relations being embedded in social relations, social relations are embedded in the market.
>
> (Polanyi, 1947: 57)

Thus, the roots of modern economics, as approaches to the age-old 'economic problem' which faces all societies, cannot be separated from the rise of the market society (and later industrial-capitalist society) and its associated theories of **political economy** (including those political economies which were critical of industrial-capitalism). As Polanyi puts it, 'Market economy implies a self-regulating system of markets; in slightly more technical terms, it is an economy directed by market prices and nothing but market prices' (1947: 43). And while in the modern world the market and 'market society' would be commonplace notions and practices, in the context of the emergence of political economy in the seventeenth and eighteenth centuries, such ideas were revolutionary and resulted in the transformation of Western societies.

## The environment and the history of economic thought: classical liberal political economy

In the history of economic thought (that is, reflections on the 'economic problem' as described above, which has a longer tradition than modern economics), one thing stands out about the role of the natural environment: namely, its reduction to a set of resources to be exploited for human economic ends. From an economic point of view, the natural environment has instrumental value, that is, is useful insofar as it can be exploited in fulfilling human wants. What is problematic about this view is *not* the instrumental valuation of the natural world that economic thought expresses. Rather the problem is the dominance of this view which 'crowds out' alternative non-economic forms of valuation and ways of relating to and thinking about the natural environment. A related problem is the misperception within modern economics that there are no 'natural limits to economic growth'. Finally, there is an equally unjustified assumption at the heart of modern economic thought to the effect that natural resources, the contribution nature makes to the human economy is a 'free gift'.

One of the first to outline the modern view of the economic problem was Locke. As discussed in Chapter 2, Locke was one of the first social theorists to offer a theory of value in which human labour was central, and also gave the classical liberal defence of private property of land and environment, which was one of the main tenets of the emerging market society. What marks Locke as a political economist is that his economic views were not motivated or based upon any idea that what he was offering was an 'objective' or 'value-free' view of the human economy in relation to the natural environment. His defence of private property in the external natural world, was in part motivated by a view in which such exclusive, individual ownership would offer some protection of the individual from arbitrary interference by political authority. He also acknowledged that privately owned land was more productive than unowned land left in a 'natural state'. That is, the unowned, and thus 'untransformed' environment, was 'valueless', since all value derives from human labour. Thus alongside the 'economic' argument that owned land is more productive than unowned land, we need also to see that for Locke (and for the classical liberal tradition as a whole), the defence of private property was also premised on a defence of individual liberty. So for Locke, private ownership was not only more efficient, in economic terms able to produce more goods and services, but also a central part in the creation of a sphere of private, individual freedom from political authority. Here, the natural environment (understood principally as ownable 'land') and property relations over it, are defended on economic and political grounds. Particular ownership relations over the external environment in Locke's social thought become central in outlining the classical liberal view of society, namely the importance of private property for social order and material prosperity.

Secondly, an important aspect of this classical liberal position can be found in the logic of Locke's view of unowned land as 'unproductive' and his explicit defence of material progress as central to the development and advancement of human society. This 'progressive' element in Locke's liberalism (which had Christian aspects) is simply another example of the historical 'spirit of the age' which puts Locke in the company of other writers and thinkers such as Francis Bacon, René Descartes and Isaac Newton. This 'spirit of the age' was the belief in material progress premised on the more efficient exploitation of the natural environment, as a result of the application of science and technology.

Thirdly, given the Christian context within which Locke was writing, it is important to see how, the appropriation and transformation of the natural environment could be justified within this Christian and often Puritan context. Within the dominant Christian culture of Europe of the time, it was accepted that the world (including humans) were 'God's Creation'. It is important to realise the theoretical and practical implications of this on regulating how the natural environment was viewed, valued and used. The first thing to notice is that since God (not humans) made the environment, they have no right to destroy it. As Locke himself put it, 'Nothing was made by God for man to spoil or destroy' (*Second Treatise on Government*, IV. 31). So while the natural environment was not the property of humans in the sense that God and not humans had made it, Locke, as we saw above did develop a justification for humans to claim parts of the natural environment ('land') as private property.

Without going into the precise detail and logic of Locke's justification of private property in the natural environment, what is important to note is that he used a particular reading of Christian virtue in order to ensure that this idea of human private property in land did not offend against the dominant and politically central idea of the natural environment being part of 'God's Creation'. Passmore (1980) suggests that the way Locke (and others after him) could square the circle of radically transforming nature (on the basis of establishing private property in it), was by justifying human transformation of (and property in) the natural environment on the grounds that humans were simply 'perfecting' or 'improving' nature. Human alteration of the world could be justified on the grounds of constituting an 'improvement' of creation for the glory of God (Passmore, 1980: 28–32). In this way, the early economic view of human relations to the natural world (which was still heavily circumscribed by Christian ideals and tradition), as a potentially productive 'resource' which requires land to be transformed into private property which can then be 'improved' (i.e. exploited and used) by humans, was tied up with the whole (Christian and particularly Protestant) notion of 'progress'. And in Locke, we see that one of the consequences of 'progress' and social development is an alteration in how the natural environment is viewed and used: that is, the necessity of the external environment becoming a commodity and an economic resource, which like any other commodity could be given a money price and bought and sold on the open market.

In this way we can observe how the 'disembedding' of the 'economy' from other social spheres and regulations (religious and political), and the emergence of a distinctly 'economic' space and motive, took place. This 'commodification' of the natural environment, which was a central aspect of classical political economy (from Locke to the eighteenth century), was thus a necessary prerequisite for the rise of modern industrial capitalism, as well as being one of the key structure features of that socio-economic order up until the present day.

## Land, labour and the enclosure movement

It is as land that the natural environment was equated in the history of economic thought, up until the first stirrings of environmental problems in the 1960s. As it was with land, its transformation into a 'commodity' like any other to be bought, sold and exchanged, that much of the pre-Enlightenment period was concerned. Much of the character of pre-Enlightenment feudal Europe was based on its rural, land-based socio-economic relations (peasant–lord relations, the guild and apprenticeship system), institutions (the monarchy and aristocracy) and customary rules (such as the peasants' right of access and use, not ownership, of common land). All of this was to change both in theory with the emergence of political economic thought, and in practice with the emergence of the self-regulating market or commercial society of the eighteenth century. But what is also important to remember is the explicit link between economic theory and actual economic practice of this time. In other words, the pre-eminence of economic thought in influencing political decisions of the day, its role as the legitimating ideology or rationalisation for the tremendous and revolutionary changes in European societies, cannot be overemphasised. Just as a combination of custom, tradition, Christianity and monarchical political requirements, furnished the pre-Enlightenment, feudal social order with its legitimating self-understanding, so economic thought functioned in a similar way to justify, defend and legitimate the radically different social and economic order of industrial capitalism. And unlike the later attempt to turn economic thought into an objective 'science' (the idea of 'positive economics'), the early history of economic thought is made up of competing schools of 'political economy' all of which share the same object of study: the political and economic basis of industrial capitalism. Political economy was *the* social theory of this historically unprecedented form of society. It reduced the seeming chaos of this emerging, bustling industrial-capitalist system, the trading, factory system, international trade and exchange, the growth of unknown levels of material wealth alongside great poverty, the urbanisation of society and the decline of the rural character of society, to its barest essentials. And in the works of classical political economy from Locke to John Stuart Mill we can find the clearest expressions of social theorising about this unique, endlessly changing and dynamic form of society.

One of the key ways in which economics captures the character of early industrial capitalism as well as providing an economic reason for creating that character, is in its attitude to the land. For the market system to work, land, labour and capital had to be 'freed' or 'disembedded' from non-economic restrictions, customs and rules. In short, the industrial-capitalist system required that land, labour and capital be 'free' to move where they were economically required, and where the market dictated they should go. Another way of putting this is that these 'resources' in order to be resources in the sense the new economic system required, had to be seen as commodities, things that could be bought, sold and exchanged. From the early 'economic' point of view, the natural world was simply a set of resources, and the most important of these resources was land. However, the problem was that land in the pre-Enlightenment period was not viewed as an 'economic resource'. Rather what we now call 'real estate' land as sellable and exchangeable property for money, was something alien to feudal, pre-industrial-capitalist society. As Polanyi puts it, 'What we call land is an element of nature inextricably interwoven with man's institutions. To isolate it and form a market out of it was perhaps the weirdest of all the undertakings of our ancestors' (Polanyi, 1947: 178). He goes on to point out that traditionally land and labour form part of the same whole, one is inextricably bound to the other.

The economic function is but one of many vital functions of land. It invests man's life with stability; *it is the site of his habitation; it is a condition of his physical safety; it is the landscape and the seasons.* We might as well imagine his being born without hands and feet as carrying on life without land. And yet to separate land from man and organize society in such a way as to satisfy the requirement of a real-estate market was a vital part of the utopian concept of a market economy.

(Polanyi, 1947: 178; emphasis added)

Thus, according to Polanyi, in the feudal world, 'land' was not viewed the way it is understood in the modern age. Land was 'home', 'place' (as opposed to just physical space or resource). In short, 'land' was the milieu of everyday life in feudal thought and experience. How did this new idea of land as an economic resource to be bought and sold on the market arise? While an incredibly complex issue, one of the key ways in which land was made into an economic resource was by dissolving the cultural and social context within which it was embedded, a major part of which was sketched in the previous section on Locke and early classical liberal political economy. Within the pre-Enlightenment social world, 'land' was not an economic resource in the required sense. This is not to say that it was not used in ways that we would say are recognisably 'economic'. Rather, and this expresses the character of this pre-modern, feudal social order, the economic functions and uses of the land were enmeshed in a whole series of cultural, social, political and religious rules and customs. That is, the idea of land as a distinctly 'economic' resource was unintelligible in the pre-capitalist, feudal era.

In England for example, while peasants did not own the land, as commoners they had customary rights of access to use common land. What happens in England is that the emergence of 'land' (a particular understanding of the natural environment) in its modern economic sense of sellable, private 'real estate' emerges after a long and bitterly resisted process of 'enclosures' which over a period of two hundred years or so (from the seventeenth to the mid nineteenth centuries) transforms the commons from a part of the cultural fabric of rural life, into private property and the exclusive economic resource of the landowner. To use an inexact modern expression, the enclosure of the commons was one of the first acts of *privatisation* of previously 'publicly' shared (if not commonly owned) property.

The enclosures were viewed as a necessary step to take in order for social development, progress and civilisation to proceed. The removal of people from the land, enabling landowners to 'develop' it and thus secure a greater return on their investment, not only was part and parcel of the birth-pains of removing the 'land' from tradition and custom, but also in the process created landless peasants who would form the urban working class in the rapidly urbanising areas. At the same time, the 'guild system' and craft-based production were eventually eroded and replaced by the modern factory system, centralised and hierarchical forms of production in which the 'worker' had little direct say in what was made or how it was produced.

Yet, as the history of the enclosures in Britain demonstrates, this radical change of a whole way of life, its set of social, cultural and political institutions, the transformation of land and labour into exchangeable, marketable resources or commodities, did not occur either overnight or peacefully. While such changes to how the natural environment was viewed and used, the creation of new 'economic' modes of acting and behaving towards both land and labour were seen as necessary for social progress; for the majority of the people at the time, these and other changes were 'costs' and they were the main 'losers' from this unprecedented erosion and destruction of a settled, familiar culture and way of life. And faced with such changes, the period leading up to and including the industrial revolution was marked by popular struggles, uprisings and resistance to the creation and maintenance of the emerging industrial-capitalist social system, such as movements like the Diggers and the Levellers.

The important point here is to note how different ideas of the natural environment and land, played a central part in the transition from the feudal to the modern industrial-capitalist socio-economic order. The main struggles concerned competing ideas of the land as 'private property', a commodity to be bought and sold, versus the claims of those for whom the land was not just a public resource, but also 'home' and an essential part of the social fabric.

## Material progress, poverty and economics

What marks the political economy of the time as a form of social theory is the explicit recognition that the changes within European societies, like all forms of societal change, would produce 'winners' and 'losers'. Much of the political upheaval associated with the emergence of the industrial-capitalist system can be seen as struggles between winners and losers. Landless, homeless peasants having been removed from the land, and forced into the cities with their labour-saving technologies, fought long and hard against the bright new world of early industrial capitalism. These losers smashed machinery (the Luddites discussed in Chapter 2), organised themselves against the emerging modern, urban economy, and, at least initially, sought to return to the security and relative prosperity of their previous rural and feudal way of life. To the economic 'spirit of the age' such disturbances and the poverty, insecurity and degradation which occasioned them, were a necessary 'price' to be paid for the new capitalist social order. As Heilbroner explicitly notes, 'The market system with its essential components of land, labor, and capital was born in agony – an agony that began in the thirteenth century and did not run its course until well into the nineteenth' (1967: 30).

Some early social theorists of industrial capitalism, like Jeremy Bentham were explicit. In his book *Principles of Civil Code* he wrote, 'In the highest stages of social prosperity the great mass of the citizens will most probably possess few other resources than their daily labor, and consequently will always be near to indigence' (quoted in Polanyi, 1947: 117). This is a theme also found in Malthus who argued that the 'poor' (the urban working class) ought to be kept at subsistence wages, since higher wages will only encourage greater increases in population of the 'lower orders' and lead to a catastrophic imbalance between human population levels and agricultural food production. Hence the 'naturalism' so typical of other ideological defenders of the laissez-faire capitalist system, such as Herbert Spencer and the Social Darwinists (discussed in Chapter 3), had a history in other 'liberal' strands of social thought of the late eighteenth and early twentieth centuries. By naturalism was meant that the market system was not only the *natural* outcome of human behaviour, but also conveyed the claim that this system *naturally* required poverty as part of its own unalterable '*natural* order'.

The link between progress and poverty was for many in the eighteenth and nineteenth centuries regarded as perfectly 'natural', and indeed as Polanyi puts it, 'Poverty was nature living in society' (1947: 84). Thus was the glaring paradox between the unprecedented productive potential of industrial capitalism in creating the most materially affluent societies the world has ever seen alongside the persistence (and indeed deepening) of poverty within such materially affluent societies. For those who either experienced this poverty and its consequences (primarily the urban working class), or who were horrified at the social and environmental harms that industrial capitalism had caused (such as the Romantics discussed in Chapter 2), a return to the material security, and relative environmental harmony of the rural, feudal social order was an attractive proposition. For many critics of the new capitalist social order, it was evident that it could not eradicate poverty, since the economic system required the poverty of the mass of its population in order for them to engage in wage-labour, which for the majority of the newly formed working class was an alien concept and one which they resisted. As Polanyi (1947) explains, although as the industrial revolution and the development of capitalism continued more people were (on some measurements, such as health and not simply financially) better off than before, there were serious costs and socio-economic changes which made people worse off.

> In spite of exploitation, he might have been financially better off than before. *But a principle quite unfavorable to the individual and general happiness was working havoc with his social environment, his neighborhood, his standing in the community, his craft; in a word, with those relationships to nature and man in which his economic existence was formerly embedded.* The Industrial Revolution was causing a social dislocation of stupendous proportions, and the problem of poverty was merely the economic aspect of this event.
>
> (Polanyi, 1947: 129; emphasis added)

In this way, the struggles against the industrial-capitalist order, which often were tied directly or indirectly with the 'defence of nature' or which advocated a 'back to the land' alternative to the urban, industrial-capitalist society, are important features of the conflict between different understandings and aspirations advanced by different classes, writers and activists concerning the relationship between 'environment', social progress and social order. At such times of crises, uncertainty and conflict over social development and change, appeals for a return to a more secure, rural and traditional past were common.

It is interesting to note that such 'back to the land' alternatives, which are an important forerunner of some contemporary 'green' alternatives to 'late' capitalism, continued right throughout the nineteenth century (Gould, 1988) and can also be observed in the mid-war period throughout Europe (Bramwell, 1989). *The important point to note is how in political struggles between defenders of industrial capitalism and those critical of it and proposing an alternative, particular constructions of 'environment' and 'nature' as well as justifications for particular sorts of relations to the natural environment are central parts of these political (and cultural) struggles.* The historical argument here thus complements the discussion outlined in Chapter 3 dealing with the different conceptions of and ideological uses of 'nature', 'natural' and 'environment' between classical liberalism (such as Social Darwinism) and critiques of liberal capitalism (such as communist or left-wing anarchism).

The next section will discuss other ways in which conceptions of 'nature' and the 'environment' were (and are) used in political economic thought to justify or establish particular arguments, positions or practices.

## Economic theory, science and environmental policy-making

In keeping with the general dominance of orthodox economic thought over public policy-making in liberal democracies, it comes as no surprise to see its centrality in environmental policy-making. Of all forms of human thought, it is economics which almost since its birth as 'political economy' and its later transformation into 'positive economics' at the end of the last century that has had the most lasting effect and hold upon political decision-making. From early forms of political economy, such as English mercantilism and French physiocracy, there has been a close, if not symbiotic, connection between 'orthodox' economic thought (i.e. broadly in support of the industrial-capitalist economy and its political requirements), and the nation-state, its institutions and decision-making processes. Unlike 'speculative philosophy', tradition, custom or religion, which had previously been the main sources of knowledge used by political authorities, economics was always regarded as a 'practical science' perfectly suited to the *realpolitik* of statecraft and political decision-making.

Given its explicit recognition of the disparity between human wants and limited means to fulfil them, economics has always presented itself, and has been perceived as uniquely equipped to deal with 'tough choices', to inform decision-making on mutually exclusive outcomes. And as it moved from political economy to economics, economic thought has sought to refine its scientific character and present itself as the objective, dispassionate study of the 'economic problem' within the context of modern, complex capitalist societies. Even anti-capitalist political economies, such as Marxism, shared the character of being 'scientific', as discussed in Chapter 3.

Now, with regard to the 'spirit of the modern age' which has been a constant point of reference so far, to be 'scientific' was (and largely still is) to be considered progressive, as well as trustworthy, rigorous, objective and dispassionate. And it is in this privilege accorded to science and the 'scientific method' that we can find roots of the dominance of economic forms of reasoning and thinking within contemporary capitalist liberal democratic nation-states. Its predominance as the central form of knowledge (along with natural science) used by state actors, bureaucracies and leaders to make decisions, implement policies and propose reforms, while of course not excluding other forms of knowledge and political judgement, has had a profound effect on the environment and social-environmental relations. As Francis Bacon, one of the founders of the modern scientific method and worldview, noted 'Knowledge is power', and this is particularly true in respect of the natural sciences and economic science. At the same time, economic forms of thinking do not simply express themselves within state policy-making, but can also seep into 'ordinary' or 'commonsense' modes of thinking. While the powerful effect of economic reasoning on modern perceptions of the environment and its official (and unofficial) influence on state decision-making which affects the environment, cannot be overestimated, I will limit my discussion to a few salient points.

The first and most obvious point is that economics as a self-styled 'science' modelled itself on the physical sciences, particularly physics and mathematics in the late nineteenth century. While it was a 'social science', its rigorous scientific methodology made it a 'hard' rather than a 'soft' form of theoretical inquiry. What is meant by 'hard' and 'soft' is that economics claimed, like the physical sciences, to be able to explain, predict and measure its subject matter, whereas the 'soft' forms of social inquiry, philosophy, sociology, politics and history, could at best 'interpret' and give meaning to rather than describe and predict causal relationships with the accuracy of economic methodology. Thus economics became (and even today is often known as) 'the queen of the social sciences'. While it borrowed the scientific method of inquiry, it also absorbed the instrumental view science had of the natural environment since Bacon and Descartes. As we noted in discussing Habermas in the previous chapter, for him, what marks the scientific method is that it requires an instrumental attitude towards the natural environment in order to produce knowledge we can use to better understand, explain and exploit it.

A second and equally important issue is how political debate over environmental issues within public policy is heavily influenced by economic forms of reasoning and argumentation. Precisely because of the dominance of economic considerations in public policy-making, environmental issues are often translated into 'economic' problems and courses of action pursued on the basis of the economic costs or benefits of the environmental issue in question. For example, in the classic case of environmental protection versus development, it is very often the case that environmental campaigners have to couch their case in economic terms and language as well as have an economic reason for environmental preservation. From campaigns to save the rainforests on the basis of the unknown medicinal substances or genetic knowledge that may be lost, to anti-roads protesters arguing their case on the basis of the drop in tourism or decline in town shopping, it seems that public policy-making requires participants in the policy-making process to adopt economic forms of reasoning and justification. To base one's case for environmental protection on the intrinsic value of the environmental space or landscape in question (as opposed to some economic-instrumental value it may possess) is to adopt a strategy that would be difficult to persuade or influence the environmental policy-making. In other words, there is a lot of strategic advantage in using economic forms of argumentation in advancing the case for environmental protection, since one is speaking a language politicians and policy-makers understand. As David Pearce has noted,

> politicians and their advisors are engaged in the activity of trading off environment against economic activity . . . *Defending the environment means presenting the arguments in terms of units that politicians understand* . . . adducing evidence that the environment does matter in economic terms is important, especially as the record of decision-making in the absence of such valuations is hardly encouraging for the environment.
>
> (1992: 8; emphasis added)

This is not to deny the importance of economic considerations, but simply to note how an economic approach to and understanding of social-environmental problems can (and does) 'crowd out' non-economic forms of environmental valuation and argumentation. The privileged position occupied by economics in environmental policy-making has the effect of drowning out other 'voices', other forms of reasoning, valuing and thinking about the environment.

Slightly over-exaggerating (but then one could say that exaggeration is when the truth loses its temper!), economic reasoning, methodology and forms of valuing the natural environment can be regarded as not simply the *language of power* in policy-making, but the *grammar of power*. What is meant by this is that economic theory functions as the dominant way in which environmental policy-making is debated, thought about and ultimately decided. In this sense it constitutes the very 'rules of the game' in the same way as grammar is the rules for the correct use of language. Thus those who either do not know or refuse to accept this particular grammar (such as non-economic arguments for environmental preservation) are at a severe disadvantage in trying to influence environmental policy-making within the current institutional framework. A good illustration of this last point is the success and rise in prominence of 'environmental economics' in the last decade or so.

## Economising the environment: the rise of environmental economics

Environmental economics holds that the environmental problems facing society can be solved by a suitably regulated market and using the tools and reasoning of neo-classical economics. The most prominent exponent of this view is David Pearce and his colleagues, who outlined the environmental economics approach in their widely read book, *Blueprint for a Green Economy* (which was a report commissioned by the then Conservative Secretary of State for the Environment, Chris Patten). This document and subsequent writings from this 'green' form of economics is characterised by its aim to translate environmental problems into economic problems. A central argument of this environmental economics position is that environmental problems arise because of 'market externalities', a market imperfection which means that environmental costs, such as pollution, are outside the market mechanism. Pollution thus constitutes an 'externality' because it does not have a price (or cost), and as there is no market in pollution, a price is unlikely to emerge. By viewing pollution as an economic problem and using economic techniques to calculate the costs it creates in terms of ill-health leading to less working hours, or greater costs to the health services, its negative effects on house prices and so on, environmental economics seeks to find the 'economic price' of pollution. Once this is arrived at, then it can be used in environmental policy-making, such as imposing a tax on pollution which reflects its cost. In this way, the 'externality' (a cost which is borne by society as a whole or a local community), through the imposition of a tax can be 'internalised' by the polluting industry or firm. This technique is also extended to calculating the economic benefits of environmental goods and services, viewed as 'natural capital' (Pearce *et al.*, 1989).

Now while clearly not without its merits, and it cannot be overemphasised how much of an advance environmental economics is over orthodox approaches to environmental problems, it does have some problems. One of these is that it depends on seeing environmental problems as economic ones, and also translating all values into economic costs and benefits. *It tries to deal with social-environmental problems by reducing them to economic ones: it economises the environment rather than ecologising economics.* The environment is viewed as an economic resource, providing economically important environmental goods and services, just as early economic thought commodified the land. As a basis for environmental protection therefore one has to ask whether this approach provides convincing reasons for the preservation of economically marginal environmental goods, species, ecosystems, etc. At the same time, there are many who question the whole idea of viewing the environment as 'natural capital' on the grounds that preserving natural capital is *not* the same as preserving the environment. As Holland has pointed out, 'insofar as there is a distinctively *environmental* crisis, it lies in the fact that the natural *world* is disappearing, not in the fact that natural *capital* . . . is disappearing' (1997: 127). Thus the preservation of natural capital will not necessarily lead to the preservation of particular parts of nature, ecosystems, species and landscapes. Hence if one wishes to preserve the natural environment, then arguing for it in terms of 'natural capital' may not be the best way to do this.

A third issue is that the dominance of this economistic form of environmental valuation and reasoning encourages a reductionist and atomistic approach, which is particularly unsuited to the holistic and integrated approach that is needed to deal with most social-environmental problems. Here in terms of environmental policy-making, it is not just the economistic forms of looking at the issue that is problematic, but also the hierarchical, segregated bureaucratic structure of the state institutions which make and implement environmental policy-making. As Dryzek (1987) notes, modern, centralised state institutions and the economistic forms of formulating environmental policy they use, are inappropriate to deal with environmental problems which require holistic and integrated solutions. What tends to happen, according to Dryzek, in centralised forms of environmental policy-making is a marked tendency towards 'problem displacement' and not 'problem solution'. Take pollution, for example. Solid pollution (domestic and industrial sewerage or garbage) can be transformed into water pollution (by dumping it into rivers or the sea) or air pollution (by incinerating it). But the point is that this simply 'displaces' the pollution, transforming it from one media to another, without actually solving the pollution problem.

Fourthly, economic views of environmental issues encourage a short-term perspective on environmental issues (based on the idea of 'discounting' future economic benefits or costs, i.e. a pound today is worth more than a pound next year or in ten years' time). This short-termism within economic forms of looking at environmental problems is often associated with and backed up by a belief that any future environmental problems that may be caused by present decisions, will be solved by technological developments. This 'techno-fix' view as O'Riordan (1981) terms it, means that we do not need to worry about long-term environmental effects (not just because as J.M. Keynes, one of the greatest economists of this century, said 'in the long run we're all dead'), but rather because the future will take care of itself by virtue of the continuing progress in science and technology in identifying, preventing and coping with environmental problems. While there have been examples of successful technological solutions to environmental problems, this does not offer a strong position to suggest that past successes will be repeated in the future to deal with environmental problems of which we are only dimly aware. At the same time, past experience of technological solutions to environmental problems have demonstrated that they often cause other or worse environmental problems ('displacing' rather than 'solving' the problem in Dryzek's terms above). For example, building higher smoke stacks on fossil-fuel-based power stations on the east coast of Britain may disperse the pollution higher in the atmosphere, but it also has the effect of causing **acid rain** in Scandinavian countries resulting in ecological damage to lakes, wildlife, soils and forests (Elliot, 1997: 25). There is thus a strong argument to suggest that this 'technological optimistic' underpinning of economic thought, while it cannot and should not be rejected completely, should, at the same time, not be viewed and used as a panacea for all environmental problems. Rather, at times, it has the character of a 'belief' or 'hope' rather than a firm or self-evident proposal, and one that, though it may sometimes be a necessary part of an overall solution, is very rarely a sufficient condition for arriving at one.

Finally, and perhaps most importantly, there is an in-built presupposition for continuous economic growth within most forms of economic theory, which goes against long-term environmental sustainability. In this assumption, this idea of 'progress' as increases in the production and consumption of goods and services, economics is simply articulating the more general and widespread idea of progress and the 'good life' which can be traced back to the **Enlightenment** as we saw in Chapters 1 and 2. In this way, economic thought can be regarded as the purest and clearest expression of the modern 'spirit of the age': the belief in a particular linear view of progress. The problem with this view of progress is that it has been associated with increasing levels of environmental degradation and the proliferation of environmental problems from local ones such as soil erosion to global ones such as climate change.

There is also another reason for the in-built bias towards growth within orthodox economics. This has to do with the social theory of orthodox economics which favours and requires socio-economic inequality. On the basis that it offers 'neutral' or objective analyses of the economy, neo-classical economics simply says that it provides the models and predications of how the economy will or should work, the distribution of income, wealth and so on are not its concern. This leads to a bias towards growth. As Mulberg explains, 'growth becomes a vital issue because of the lack of an adequate (or indeed any) distributional theory within mainstream economics . . . In practice economic growth has acted to deflect questions of redistribution' (1995: 147; Barry, 1999: ch. 7).

All this seems to add substance to Milton's (1996) view of the innate conservatism of economics. According to her, 'While sociology and political science (and for similar reasons, cultural anthropology) are inherently subversive, economics, at least in its neoclassical form, cannot be' (1996: 72). However, there has been a recent attempt to make economics more 'subversive', namely ecological economics to which we turn next.

## Ecological economics

Building on, though largely arising out of a critique of environmental economics, has emerged ecological economics. As Juan Martinez-Alier, one of the leading ecological economists describes it, 'The inability of orthodox economics to cope with green issues has given rise to

**Figure 6.1** *'Economic Growth'*

*Source:* Polyp, P.J., in *New Internationalist*, June, 1996

ecological economics, which is the study of the compatibility between the human economy and ecosystems over the long term' (1997: 22).

In many respects ecological economics represents a return to the tradition of *political economy*, which orthodox economics has long since abandoned in its aim of being 'objective' and 'scientific'. That is, ecological economics does not regard the solution of the 'economic problem' to be a purely 'economic' matter. Rather it recognises that the economy is not only a subsystem within a wider ecological system, but also operates within and is influenced by a wider political and cultural context. The economy *ought not* to be disembedded from politics and culture or its ecosystemic basis.

Ecological economics seeks to integrate nature's economy (ecosystems) and the human economy, which requires that the latter be seen as dependent upon and a subset within the former. As this is the main focus of ecological economics, it differs from mainstream economics in that it seeks to base its theories and models on the insights of natural science (particularly ecology and thermodynamic theory from physics) as well as having roots in economic science. In its focus on the wider social context of the economy it is an institutional form of economics (Jacobs, 1994), which together with its natural science basis, makes it an explicitly multidisciplinary approach to the 'economic problem' introduced at the start of this chapter. As van den Bergh puts it, 'ecological economics should not only search for common elements, theories and approaches in the sciences of economics and ecology. It should try and take a broad view encapsulating economic, social, ethical, historical, institutional, biological and physical elements' (1996: 35–36). Thus ecological economics requires that we develop a new set of terms, models and conceptual tools in order to accurately theorise the interaction beween ecology and economy. It expands one of the aims of orthodox economics, namely, managing the human economy to include managing the natural environment and the human economy together. Of central concern to ecological economics is the issue of the scale of the economy in relation to its ecological basis, something about which mainstream economics has little to say.

Ecological economics is an extremely new and challenging development within economic theory, and while it has succeeded in becoming a subdiscipline within economics it is still a minor school of thought. As yet it does not have the same hold on environmental decision- and policy-making as orthodox economics or environmental economics. It does, however, represent an attempt to bring together the study of the human economy (economics) and nature's economy (ecology) within one multidisciplinary body of knowledge.

## Conclusion

Of all the social sciences, economics, from its origins in political economy, has perhaps had the most effect on how the natural environment has been viewed, valued and treated in Western societies. As a form of social theory, it has had widespread and far-reaching consequences for how the relationship between society (and the economy in particular) and the natural environment has been thought about and analysed. Historically, political economy and its associated economic worldview, was used to legitimate the various changes that constituted the transition from feudalism to an industrial-capitalist market society. In the modern world, the heir of political economy, orthodox 'economics' or positive 'economic science', has a powerful hold on public policy-making, and environmental policy-making in particular.

At the same time the 'economic' view of the natural environment is one which has commonsensical appeal. Most people would go along with the economic view of the natural environment, i.e., that it has only instrumental value to humans and its instrumental value is of an economic form. This economic value of the natural environment is in terms of its functions as a 'resource' or 'input' to the human economy. Economics is thus a major form of social theory and practice which upholds an anthropocentric or human-centred view of nature and human relations to it. Its importance lies in that its area of study, i.e. the human economy, is the one part of human society which has a direct and material interaction with and relation to the natural environment. The economy is where nature and society meet.

## Summary points

- Economics is a form of social theory, indeed perhaps one of the most powerful and important in terms of how the natural environment is viewed, valued and treated.

- One of the central achievements of the emergence of an 'economic' view of society in the seventeenth and eighteenth centuries was the 'disembedding' of the economy from wider social, political, cultural and religious context.

- The emergence of liberal political economy led to the enclosure movement, the commodification of the land and a view of the natural environment as only of instrumental value to humans.

- There was a historical link between resistance against and criticism of the 'economic worldview', the rise of industrial capitalism, the nation-state system in Europe and particular traditional views of 'nature' and 'land' and rural ways of life in relation to the land as 'place'.

- There is a strong link between orthodox economic thought and policy-making in industrial societies.

- Environmental economics, in which environmental goods and services, as well as environmental risks and bads, are given a monetary value, is the first systematic attempt to introduce the environmental dimension within mainstream economics.

- Environmental economics can be criticised for economising the environment rather than ecologising economics.

- Ecological economics is a recent development within economy theory which attempts to integrate economy and ecology, that is, is explicitly based on the dependence of the human economy upon nature's economy.

- Ecological economics is a mutlidisciplinary approach to ecological-economic interaction, which not only integrates natural science into its analysis, but in its attention to the political, cultural and social context of the economy can be seen to represent a return to the older tradition of 'political economy'.

# Further reading

For some general reading on the relationship between economics and the natural environment see Donald Worster's *Nature's Economy: A History of Ecological Ideas* (Cambridge: Cambridge University Press, 1994); P. Mirowski (ed.), *Natural Images in Economic Thought: Markets Read in Tooth and Claw* (Cambridge: Cambridge University Press, 1994); and Jon Mulberg, *Social Limits to Economic Theory* (London: Routledge, 1995).

On the historical relationship between political economy and the natural environment see, A. Clayre, (ed.), *Nature and industrialization* (Oxford: Oxford University Press, 1997).

On environmental economics, see David Pearce *et al.*, *Blueprint for a Green Economy* (London: Earthscan, 1989); and the rest of the 'Blueprint' series published by Earthscan, particularly D. Pearce *et al.*, *Blueprint 3: Measuring Sustainable Development* (London: Earthscan, 1993). For critical assessments of environmental economics, see Michael Jacobs, 'The Limits of Neoclasicalism:

Towards an Institutional Environmental Economics', in M. Redclift and T. Benton (eds), *Social Theory and the Global Environment* (London: Routledge, 1994); and John Barry, 'Green Political Economy', ch. 6 of *Rethinking Green Politics* (London: Sage, 1999).

On ecological economics, see Robert Constanza (ed.), *Ecological Economics: The Science and Management of Sustainability* (New York: Columbia University Press, 1991); Juan Martinez-Alier, *Ecological Economics* (Oxford: Basil Blackwell, 1987); Nicholas Georgescu-Roegen, *The Entropy Law and the Economic Process* (Cambridge, MA: Harvard University Press, 1971).

# Risk, environment and postmodernism

- Beck's 'risk society' thesis
- The character of risk
- The 'precautionary principle'
- Reflexive modernisation and the redefinition of 'progress'
- Democracy, democratisation and risk society
- Postmodernism, social theory and environment
- Postmodernism, environmentalism and the rejection of modernity
- Postmodernism and post-industrialism
- Postmodernism and the social construction of the environment
- Problems of postmodern environmentalism

## Introduction

Throughout the book so far an important point of reference for analysing the environment and social theory has been the ways in which, both historically and conceptually, 'modernity' has affected social theorising about the environment. '**Modernity**' can be understood as the (sometimes radical) changes in the organisation and legitimation of 'modern' social, political and economic life and 'modern' ways of thinking and acting, associated with the advent of the modern' age in the latter half of the eighteenth century in Europe. The many and complex aspects of these changes in almost all parts of life are central for an adequate understanding of how 'social theory' (itself a product of the 'modern' age), in its different forms, schools and as developed by different social theorists, viewed, valued and thought about the environment.

In this chapter, we discuss two contemporary forms of social theorising about the environment and the environmental crisis, to which 'modernity' and its legacy are central to both. These are the 'risk society' thesis associated with Ulrich Beck, and the relationship between postmodern social theory and the environment. Both can be seen, in sometimes very different ways, to be directly engaging with the problems and potentials of modernity and its legacy.

In the first part of the chapter, Beck's 'risk society' thesis is explored which in terms of its attitude of modernity, can be broadly viewed as representing an 'immanent critique and reconstruction' of modernity. That is, 'risk society' can be seen as an attempt, in Beck's own terms as a 'new modernity' (which is the subtitle of his famous book, *Risk Society*, published in English in 1992). Thus while Beck does criticise modernity, particularly in terms of the rise in potentially catastrophic environmental risks, he does suggest that modernity has within itself the ability to solve the problems it produces.

The second half of the chapter looks at postmodern social theory and its treatment of the environment and associated phenomenon such as the environmental crisis and environmental politics. Postmodern social theory as its name suggests, unlike Beck's social theory, sees itself as 'beyond modernity'. While no such thing as *a* singular or homogeneous postmodern social theory exists (there are many varieties of postmodern social theory), what the different varieties of postmodernism share is an extremely critical and radical assessment of 'modernity'. In terms of 'modern' forms of knowledge (from natural science to social thought), and modern political, social and cultural institutions (such as the nation-state, liberal democracy, the nuclear family), to 'modern' alternatives to liberal capitalism (particularly Marxist ideology), postmodernism suggests that the various problems associated with modernity (including environmental ones) cannot be solved within modernity, but require a postmodern solution.

## Beck's 'risk society' thesis

Stated bluntly, Ulrich Beck's 'risk society' thesis suggests that what we are witnessing in contemporary Western societies is the emergence of a politics concerned with the interpretation and distribution of social and ecological 'bads' rather than 'goods'. The politics which characterised 'industrial society' centred around the production and distribution of 'goods' such as wealth, income and formal employment. As Beck puts it, 'What was at stake in the older industrial conflict of labor against capital were positives: profits, prosperity, consumer goods. In the new ecological conflict, on the other hand, what is at stake are negatives: losses, devastation, threats' (Beck, 1995a: 3). For Beck, 'risk society' refers to the recent transformation of Western societies, with specific focus on environmental issues. It concerns the health, socio-economic, cultural and environmental effects of 'social progress' in general, and scientific and technologically based production in particular.

In industrial society, politics is largely concerned with the distribution of the benefits of society (wealth, income, jobs), hence the dominant political conflict is between 'capital' and 'labour', and political life assumes a 'left–right' character. In risk society, which is located between 'industrial' and 'post-industrial' stages of social advancement, the dominant focus of politics is the distribution of the 'costs' and 'risks' of socio-economic development, dominated by the emergence of unexpected ecological and health hazards.

In this new historical epoch, politics, to use a famous green slogan, is 'beyond left and right'. Beck's ecologically informed analysis of risk and his account of the emergence of a 'post' left–right political landscape is something shared with Giddens, discussed in Chapter 4. Also in common with Giddens (and Habermas), Beck's 'risk society' thesis has both descriptive and prescriptive dimensions. While he is *describing* recent social changes in Western societies, he also *prescribes* ways of dealing with these changes and promotes a particular vision of an 'ecologically rational' or 'ecologically enlightened' society.

There is some empirical evidence to support Beck's thesis that Western societies are (or are becoming) 'risk societies', both in terms of greater awareness and sensitivity of risk, and 'at risk' from their own development paths. In 1995 a MORI poll in the United Kingdom asked the following question: 'Do you think that the kind of world that today's children will inherit will be better or worse than the kind of world that children of your generation inherited, or about the same?' (Jacobs, 1996: 3). The results were:

| | |
|---|---|
| Better | 12% |
| Worse | 60% |
| About the same | 25% |
| Don't know | 4% |

What are the reasons for this extremely pessimistic assessment of the future? Why do people think the future will not be better than today? The number and range of possible risks or bads we can think of include: the increase in crime; a decrease in personal safety; rising unemployment levels; declining job security and prospects; the breakdown of families and communities; the risk of contracting diseases/illnesses; the public controversies about the decline in food safety; and rising pollution levels and environmental degradation. Across these and other issues, one can witness the growing risk sensitivity of Western publics due to experience of one 'scare' or 'risk' after another: such as the 'mad cow' disease (BSE) in the United Kingdom, the outbreak of E Coli food poisoning, as well as environmental dangers such as global warming and ozone depletion, and environmental-related illnesses such as the dramatic rise in childhood asthma, linked to the increase in car pollution.

Another form of empirical evidence of the 'riskiness' of modern societies is the rise in the quantity and quality of insurance cover that individuals, groups, firms and corporations have taken out to insure against the various forms of risk outlined above. For example, the risk of global warming, its potentially damaging effects on agriculture, water generation and distribution, and the likely devastating effects of rising sea levels for low-lying human settlements can be observed in the quantity of 'global-warming'-related insurance being taken out by companies and being held or insisted upon by the insurance industry. Or observe the rise in the private health insurance market in Britain in recent years (though this may be explained in part as a consequence of the decline of the National Health Service, as much as an increase in peoples' sensitivity to health-related risks).

In Beck's terms, these 'risks', 'bads' or 'dangers' are the side-effects (the costs) of the particular 'development path' or type of modernisation which characterises modern societies. Basically, his point is that the costs of modernisation are beginning to outweigh the benefits. As Beck puts it, '"Risk society" means an epoch in which the dark sides of progress increasingly come to dominate social debate. What no one saw and no one wanted – *self endangerment and the devastation of nature – is becoming a motor force of history*' (1995b: 2; emphasis added).

## The character of risk

We can divide risks into a number of categories.

1 *Ecological risks*: global warming, biodiversity loss, ozone depletion, ecosystem destruction.

2 *Health risks*: health risks due to genetically altered food-stuffs, skin
cancer, food safety scares ('mad cow' disease), pollution-related
illness such as asthma, cancer, heart disease.
3 *Economic risks*: unemployment and decline in job security.
4 *Social risks*: decline in personal safety, rise in crime and breakdown of
community, rise in divorce and separation.

The character of risks is important in understanding them. Firstly, they
are often 'intangible' and remote. Secondly, they are characterised by
uncertainty and unpredictability which makes them difficult to insure
against. Thirdly, they are often unknown, that is those who are subject to
specific risks are often ignorant of them. Yet there is, according to Beck's
thesis, a growing fear from risks in general. Due to both of these issues
(uncertainty and knowledge/ignorance of risks) of crucial importance in
analysing risks is knowledge and its communication which identify,
inform and define the shape and content of risks. Of particular
significance for Beck are science and the media, and one of the main
aspects of his thesis is the increasing gulf between 'science and
technology' and 'society' in general, and the latter's distrust of the
former in particular.

## Risk, trust and science

According to Beck's 'risk society' analysis, science (and technology)
are increasingly seen as the causes of modern environmental (and other)
risks rather than the solution. What marks 'risk society' from the
previous 'industrial' stage of social development is that whereas in the
latter science and technology were seen as positive forces for social
progress, a view common to nineteenth-century social theory as
discussed in Chapter 2, in 'risk society' this equation of scientific and
technological advancement and social progress is broken. Risk society
describes a modern sense of fear, distrust and unease about scientific
and technological developments. However, for Beck this distrust is not
confined to science and technology, but can also be seen in the erosion
of 'trust' in dominant social and political institutions, such as industry
and government. Here Beck's 'risk society' argument complements
other theories which analyse the 'legitimation crisis' of Western
societies, such as the early work of Habermas (1975), and other
neo-Marxist social theorists such as Claus Offe (1984).

Apologies to Walt Disney

**Figure 7.1** *'Be Very Afraid'*
*Source:* Peter Clarke, *The Guardian*, 29 May, 1996

Where once people had unquestioning faith and belief in scientific knowledge and technological developments, the experience of the negative side-effects of science and technology has led Western publics to be less trusting of 'the scientific establishment'.

This development, however, does not mean the simple rejection of science, for as Beck suggests, 'public risk consciousness and risk conflicts will lead to forms of scientization of the protests against science' (Beck, 1992a: 161). What he means by this is that scientific knowledge is no longer a unified body of knowledge, but fragmented, and 'risk society' implies the break-up of a homogeneous understanding of 'science' as speaking with one voice. It terms of the association between expert knowledge and power, it is important to stress that the authority and legitimacy accorded to scientific knowledge was in part related to its unified, internally coherent character. Once it begins to fragment and no longer be a unified whole, its ability to command automatic authority and public trust begins to wane.

Firstly, in various environmental controversies, which rely on 'scientific evidence' both to 'identify' the problem as well as suggest solutions, we increasingly find government or industry scientists facing those of environmental **NGOs (non-government organisations)**, such as Greenpeace, in presenting scientific evidence to establish or prove that a particular environmental problem exists and assess its severity. Indeed, there is some evidence to suggest that the public distrusts government or business scientists and scientific evidence and has more trust in the scientific arguments of environmental NGOs.

Secondly, there are issues on which science itself is divided, such as certain illnesses and ecological risks such as global warming (though this links with the previous point in that those scientists who dispute that global warming is due to human carbon emissions, who are in a minority, are largely funded by what is known by the 'fossil fuel' lobby, corporations such as car manufacturers, and the oil industry, who clearly have a lot to gain from discrediting global warming). Beck sees this pluralisation or breaking-up of a unitary science as something positive. The reason for this is that, 'Only when medicine opposes medicine, nuclear physics opposes nuclear physics . . . can the future that is being brewed up in the test-tube become intelligible and evaluable for the outside world' (Beck, 1992a: 234). In other words, having as many scientific voices as possible debating amongst themselves and with the public (and thereby educating the public) is the best way to identify and possibly cope with or solve the various ecological and health risks that face modern societies.

Developments in science and technology have profoundly affected their relationship with society and between society and nature according to

Beck. On the one hand, scientific and technological innovations have turned society into a laboratory. The proposed development and release of genetically modified organisms into the environment and food chain, or the 'cloning' of organisms, with their unknown effects, can be taken as an example of how, in Beck's terms, we live in an 'experimental' society. Society is being subjected to experiments over which it has no direct control and often unknown to it.

On the other hand, the large-scale, global effects of advanced scientific and technological development has resulted in what McKibben (1989) has called the 'Death of Nature'. What is meant by this is not the death of the natural world as such, but rather the end of nature as an independent entity from human intention and activity. With the advent of human-induced global warming and climate change, as a negative and unintended 'side-effect' of human productive activity, we have what can be called the 'humanisation of nature'. However, unlike the humanisation of nature that Marx advocated, the current process of the humanisation of nature is unintended, undemocratic and not based on the principles he advocated. Indeed, it is truer to say that what we are witnessing is not the humanisation of nature, but the 'capitalisation of nature' (Barry, 1995c, 1998c). Global warming and climate are largely the result of globalised processes of capitalist production, while at the same time, it is not the human species as a whole which is to blame, but disproportionately the advanced capitalist nations of the world, and large corporate multinationals.

On the other, technological developments and innovations such as biotechnology, especially genetic engineering, cloning and the creation of genetically modified organisms, are other examples of the 'experimental society'. For some, these parts of the experimental society are something to be feared. As Levidow and Tait put it, 'In parts of the popular imagination, biotechnology is feared as a violation of nature, which may then go out of control' (1995: 123). While for some people risks are exciting challenges to be overcome, and there are those who positively enjoy taking risks, for most people risks are to be feared and avoided if at all possible.

## The 'precautionary principle'

While Beck does not directly talk of the precautionary principle, it is clearly consistent with the main thrust of his thesis, and constitutes an

**Figure 7.2** *'We're All Gonna Die'*
Source: Poly, P.J., *New Internationalist*, September, 1996

important aspect of the relationship between social theory and environmental risks. According to Beck, 'insisting on the purity of scientific analysis leads to the pollution and contamination of air, foodstuffs, water, soil, plants, animals and people' (1992a: 62). In other words, in face of uncertainty and high levels of scientific disagreement, waiting to 'scientifically prove' a particular ecological risk (for example, the link between cancer and nuclear power plants) means the burden of proof is on those who wish to prevent or stop some industrial process, rather than on those who promote it. Meanwhile the negative consequences of the process continues, i.e. people still get cancer.

The precautionary principle, which has come to be a central principle of environmental risk assessment and management in recent years, holds that in the context of uncertainty it is rational to be prudent and not proceed with a particular action if there is a risk of it resulting in future significant danger or harm. For O'Riordan and Jordan,

> At the core of the precautionary principle is the intuitively simple idea that decision makers should act in advance of scientific certainty to protect the environment (and with it the well-being interests of future generations) from incurring harm . . . In essence it requires that risk avoidance become an established decision norm where there is reasonable uncertainty regarding possible environmental damage or social deprivation arising out of a proposed course of action.
>
> (1994: 194)

Core aspects of the precautionary principle include a willingness to act in advance of scientific proof and, for example, to deliberately hold back from certain forms of development on precautionary grounds. It also implies a reassessment of the belief that technology can solve all environmental problems, a belief, which as indicated above and in Chapter 2, was typical of social beliefs and social theory of 'industrial society'. Following on from these two, the precautionary principle seeks to avoid irreversible forms of environmental intervention. That is, the precautionary principle suggests that large-scale developments which cannot be later reversed (such as destroying a particular ecosystem or perhaps damming a river), and which are risky should not proceed, as the costs would outweigh the benefits. There is also a deeply normative dimension to the precautionary principle in that, in seeking to prevent certain forms of 'risky' development, it raises fundamental issues about the direction of social development and the type of society we wish to live in.

As such, the precautionary principle fits with Beck's view that environmental risks and dilemmas are not simply technical 'problems' to be 'solved' by scientific and/or economic applications. Rather, in 'risk society' the precautionary principle acknowledges the normative character of environmental risks – that is, they are moral questions about right and wrong and not simply about costs and benefits or technical 'problems' and 'solutions'. The normative character of the precautionary principle can be seen if dealing with environmental risks is viewed in the context of avoiding unnecessary harm to future generations and the nonhuman world. In this way, the precautionary principle suggests a shift from a 'polluter pays' to a 'polluter proves' principle. That is, those proposing large-scale developments with large and unknown or unquantifiable environmental effects, or developments resulting in irreversible environmental degradation (such as dams, road-building schemes, the creation, release and use of genetically modified organisms), should demonstrate that no long-term or irreversible ecological, health or other harms will result. In conclusion, the precautionary principle is in many respects simply common-sense prudence and caution, a virtue all the more pressing in the complex, uncertain world of contemporary social-environmental relations. The spirit of the precautionary principle is nicely expressed by St Thomas Aquinas's wise opinion that 'It is better for a blind horse that it is slow'.

The effects of a widespread application of the precautionary principle would be potentially radical as it would result in a qualitatively different form of development from the standard 'modernisation' model. At the very least it would mean that a stricter and political regulation of economic development, in general, and in particular, forms of modernisation which either rest on innovations and applications of science and technology and/or require the 'use' or consumption of the nonhuman environment. This different form of ecologically informed and sensitive modernisation, consistent with Beck's 'risk society' position, is discussed next.

## Reflexive modernisation and the redefinition of 'progress'

For Beck what may be called 'simple' or 'industrial/economic' modernisation is the form or model of social development associated with the standard 'Western' view of progress and development. This modernisation model is associated with the historical evolution of 'Western' societies over the last two hundred years. Some of the main features of this modernisation model include: the industrialisation of the economy; urbanisation and the emergence of a 'post-agricultural' commercial society; the creation of a nation-state; and above all else the equation of 'progress' with continuous increases in the production and consumption of material goods and services. With the collapse of communism, this 'Western' model has in recent years evolved to include the importance of 'free markets', private property and international free trade. For Jacobs this 'dominant model' of progress can be understood as meaning that

> the principal purpose of economic activity is to raise incomes. Income growth makes people better off: it enables them to consume greater quantities of both material and non-material goods, and through taxation enables governments to provide essential public services such as education, health care and social security. Given technological improvements to productivity, annual economic growth not only generates jobs, but is required to sustain them. The motor of growth is free trade: as import duties, foreign exchange controls and other forms of protection are lowered, goods, capital and labour flow to where they are most productive, and more wealth is generated . . . free trade and growth lead to 'modernisation' – the increasing productivity of agriculture, the movement of people into towns and cities and the transformation from traditional to modern cultures.

(1996: 8)

Now, for Beck, the advent of 'risk society' marks the threshold beyond which ecological and other risks outweigh the benefits of further economic growth associated with the model of industrial or simple modernisation. According to Beck, what is now required is a qualitatively new model of modernisation, what he calls 'reflexive modernisation'. The basics of reflexive modernisation are that, 'Modernisation *within* the path of industrial society is being replaced by a modernisation of the *principles* of industrial society' (Beck, 1992a: 10). As the term 'reflexive' implies, what Beck (in agreement with Giddens who also focuses on the 'reflexivity' of social institutions) means is that modernisation should mean that society as a whole increasingly reflects upon its own development and the institutions which further and/or realise that development. It is important to stress that Beck is *not* arguing from a 'postmodern' perspective, but is arguing instead for a new type of modernisation. This new model of modernisation, reflexive modernisation, can be seen as a form of social learning, an attempt by society, through its political institutions (especially through democratic politics), to deal with and/or cope with the ecological and other risks and dilemmas that have arisen as a result of simple or industrial modernisation.

A central part of reflexive modernisation is the redefinition of 'progress'. According to Beck, '"Progress" can be understood as legitimate social change without democratic political legitimation' (1992a: 214). In this way, 'progress' as orthodox economic growth (in accordance with the 'standard' or 'dominant' model) is giving way, in Beck's analysis, to 'social progress' (1992a: 203). By the latter, Beck means *institutionalised self-criticism* (reflexivity) and increased opportunities for individuals to be conscious of and deliberate upon, and democratically pass judgement on the *principles of modernisation* and not just specific policies associated with it. In other words, by reflexive modernisation, Beck means a redefinition of social progress a central aspect of which requires the separation of 'techno-economic' progress and 'social' progress. Additionally, and radically, what reflexive modernisation implies is that society democratically make decisions on its development path, that is, democratically 'regulate' social progress. The politics of 'risk society' thus concerns both the *direction* and the *substance* of social progress, and thus of social organisation as a whole.

Beck's reflexive modernisation argument holds that current ecological and other risks will only be resolved if we begin from the question, 'How do we wish to live?'. In democratically regulating social progress, what Beck is saying is that we choose to live in a different type of society, one more open to popular, democratic control and accountability. This democratisation process for Beck extends into those areas of social life, such as science and technology, which can dramatically affect people's lives, yet over which they presently have little or no control.

## Democracy, democratisation and risk society

Risk society, although it denotes social change in which job insecurity, social unease and ecological hazard figure large, thus also presents opportunities for greater democratic accountability and institutional innovation. Indeed, one can say that Beck's position is that such democratisation, the opening up of previously 'hidden' areas to democratic accountability, is not just desirable, but absolutely necessary to deal with the problems of risk society. Allied with his critique of the various ecological (and other) risks associated with industrial society, Beck also criticises the latter for its limited and limiting form of democracy. According to him, *'Industrial society has produced a "truncated democracy", in which questions of the technological change of society remain beyond the reach of political-parliamentary decision-making.* As things stand one can say "no" to techno-economic progress, but it will not change its course in any way' (Beck, 1992b: 118; emphasis added). In order to rectify this, dealing with ecological risks calls for more not less democracy, more public openness, democratic accountability and popular participation in decision-making in science and technology. The simple question Beck asks is this, 'Why should democracy end at the laboratory door?', especially since the consequences of what is decided in the laboratory are potentially far-reaching. In seeking to extend democratic control and regulation over 'techno-economic progress', Beck can be said to be simply asserting the classical defence of democracy, that is, those who are affected by decisions ought to have some say over how those decisions are made. As Levidow puts it, 'Risk assessment serves as an implicit ethics, and even as a potential means to democratize industrial development . . . Risks lie across the distinction . . . between value and fact and thus across the distinction between ethics and science' (1995: 186).

Examples of this extension of democratic accountability include more 'right to know' information about state and corporate business decisions which affect the environment and people's lives; more opportunities for citizens to influence decisions which affect them; more democratic regulation of 'unaccountable' market decisions of firms; greater popular participation in public inquiries (from proposed new road developments to genetic engineering), and more overall democratic regulation of science and technology.

Another reason for the extension of democratic norms of accountability and openness to science and technology relates to the point made above that according to the 'risk society' thesis, environmental risks are not simply technical matters to be dealt with by 'experts', but are also acknowledged as moral issues which society needs to debate and resolve. Thus, the recent controversy over genetic engineering can be seen as a classic example of what Beck means. Public disquiet and fears over cloning, genetic engineering and the creation of 'new' organisms, is in large part, motivated by a moral sense of unease with this particular form of human technology. While people are worried about the possible health and safety implications of genetic technology, there is also a deeply moral dimension to the issue, with some people thinking it is not right for humans to 'play God' in creating new forms of life, while others feel that the development of such technology indicates a violation and lack of respect for the intrinsic value of the natural world.

Beck's argument that 'risk society' requires the extension of democracy to deal with and possibly prevent environmental risks from arising in the first place (by the application of the precautionary principle for example) can be viewed in terms of the two 'imperatives' of modernity discussed in Chapter 2. Recalling the shorthand description of modernity as being made up of the French and industrial revolutions, where the former represents the democratic imperative of modernity, while the latter represents its quest for material comfort and wealth, one could say that with 'risk society' we come full circle. In many respects Beck can be said to be using the democratic imperative of the French Revolution to control the potentially negative effects of the industrial revolution and its legacy. That is, 'risk society' and its model of 'reflexive modernisation' can be seen as requiring enhanced democratic control of and public accountability of industrial progress, a central part of which leads to the democratic 'redefinition' of what constitutes progress.

Another recent version or movement within social theory, which like Beck's 'risk society' focuses on the relationship between 'modernity' and environment, is postmodernism which we turn to next.

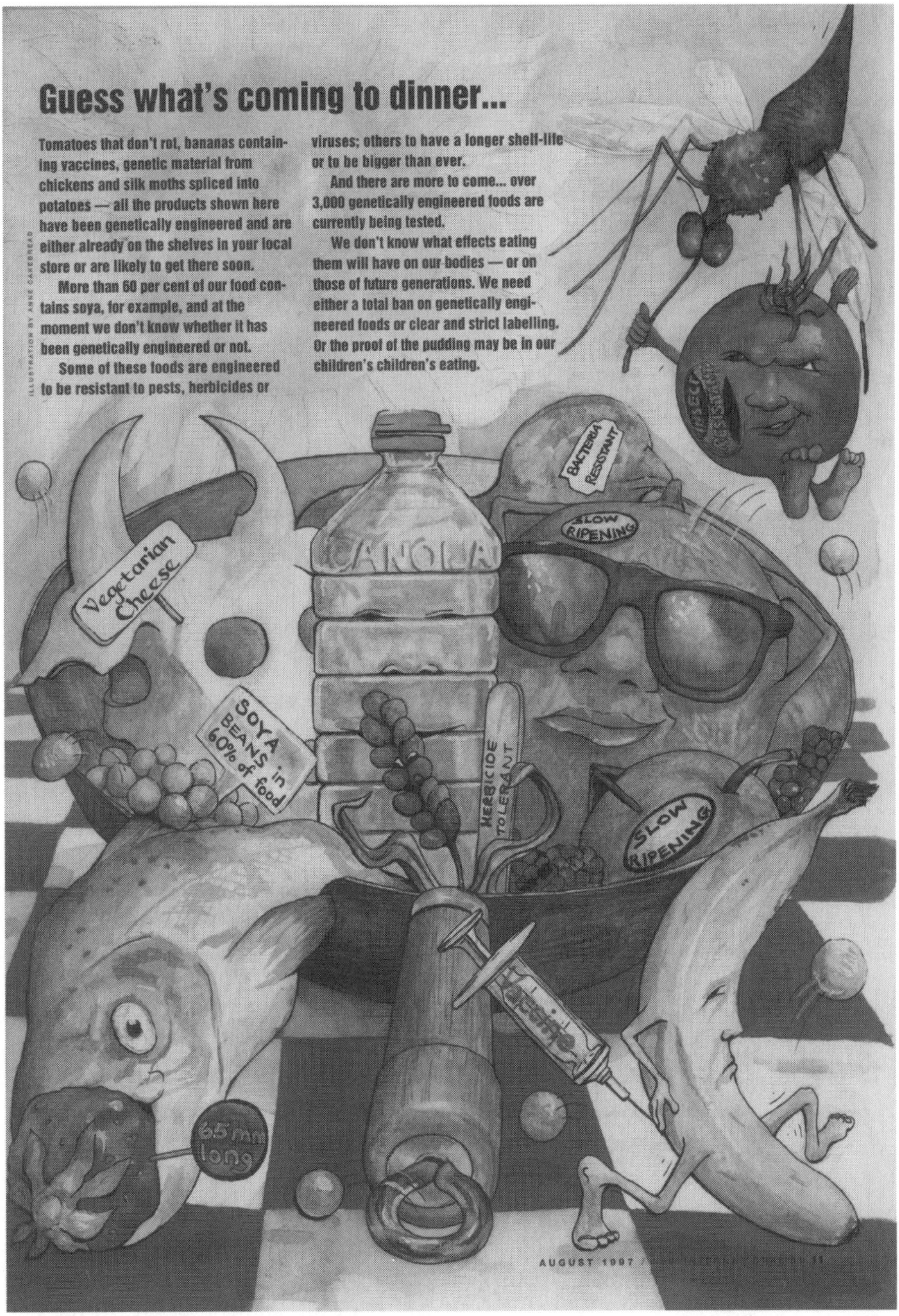
Guess what's coming to dinner...
Tomatoes that don't rot, bananas containing vaccines, genetic material from chickens and silk moths spliced into potatoes — all the products shown here have been genetically engineered and are either already on the shelves in your local store or are likely to get there soon.
More than 60 per cent of our food contains soya, for example, and at the moment we don't know whether it has been genetically engineered or not.
Some of these foods are engineered to be resistant to pests, herbicides or viruses; others to have a longer shelf-life or to be bigger than ever.
And there are more to come... over 3,000 genetically engineered foods are currently being tested.
We don't know what effects eating them will have on our bodies — or on those of future generations. We need either a total ban on genetically engineered foods or clear and strict labelling. Or the proof of the pudding may be in our children's children's eating.
Vegetarian Cheese
CANOLA
BACTERIA RESISTANT
SLOW RIPENING
SOYA BEANS in 60% of food
HERBICIDE TOLERANT
SLOW RIPENING
Vaccine
65mm long
AUGUST 1997

**Figure 7.3** ***'Guess What's Coming to Dinner'***

*Source:* Cakebread, A. *New Internationalist*, August, 1997

## Postmodernism, social theory and environment

Postmodern social theory (if one can say such a single 'school' can be said to exist) is one of the most recent developments in Western social thought, one of the many characteristics it shares with ecological or green moral and political theory. The aim of this section is to explore the extent to which environmental issues and problems can be said to be 'postmodern' is some sense, and examine the contribution of postmodern social theory to the analysis of environmental issues.

According to Docherty the term 'postmodern' 'hovers uncertainly . . . between – on the one hand – extremely complex and difficult philosophical senses, and – on the other – an extremely simplistic mediation as a nihilistic, cynical tendency in contemporary culture' (1993: 1). Postmodern social theory has many roots and crosses many disciplines, and its origins can be found in such thinkers as Friedrich Nietzsche, Martin Heidegger and Sigmund Freud. In social theory one can point to the Frankfurt School of critical theory as a key origin, particularly Horkheimer and Adorno's *Dialectic of Enlightenment*, discussed in Chapter 4.

In pointing out the dark side of 'modernity', its various costs, risks and dangers, Adorno and Horkheimer created the theoretical ground from which postmodernism began. Indeed, in the *Dialectic of Enlightenment* one can also see the connection between modernity/modernisation, postmodernity and ecological concerns, as evident in Adorno and Horkheimer's statement that 'The fully enlightened earth radiates disaster triumphant' and 'What men want to learn from nature is how to use it in order to dominate it and other men' (1973: 3).

Postmodernism is often associated with the abandoning of 'grand narratives' (Lyotard, 1984), such as 'progress', or at least the Western views of industrial and economic progress discussed in Chapters 2 and 6. Other 'grand narratives' which postmodernism rejects or suspects include Marxism and its 'myth' of a worldwide post-capitalist, communist society. Postmodernism also rejects or is sceptical of the grand narratives of science and scientific forms of inquiry which purport to reveal the 'truth' about the human or nonhuman worlds. It also focuses on the self, the individual subject, both the creation of this 'social category' within modernity as well as its problematic status. At the same time some have said that postmodernism is more concerned with the 'body' than with the 'body politic'. The latter is particularly evident in the work of Foucault. Most postmodernists see themselves as against 'totalising' (and therefore potentially oppressive) forms of discourse and power and celebrate diversity, tolerance, difference, fragmentation and self-realisation. Above all else however, postmodern social theory takes a social constructionist approach, seeing 'nature' and the 'environment' for example as socially constructed categories, brought into being by the operation of discourses and power.

## Postmodernism, environmentalism and the rejection of modernity

According to Gare, 'It is the idea of environmental crisis as "Enlightenment gone wrong" which has encouraged the view that postmodernism represents a solution to the environmental crisis through a rejection of the modernist project' (1995: 5; Gandy, 1996: 26). Eder, in similar terms, declares that 'Modernity represents a culture which has reduced the material appropriation of nature to the exploitation of nature and thus provoked the ecological crisis' (Eder, 1996: 23). In simple terms, if the ecological crisis is caused by 'modernity' and its legacy, then the solution must be 'post-modernity' something other than modernity as we know it. The ecological crisis as a 'side-effect' of modern societies and the underlying logic and practices of the 'modernist project' thus forms a strong link between postmodern social theory and the environment. As leading green theorist and activist Charlene Spretnak put it, 'Green politics goes beyond not only the anthropocentric assumptions of humanism but also the broader constellation of values that constitute modernity' (1996: 533).

Postmodernism's critique of 'scientific rationality' and the scientific paradigm (part of its wider critique of Enlightenment reason and rationality) is similar to and overlaps with radical green critiques of how science has been central in 'disenchanting' nature and transforming it into a collection of resources, whose value is determined instrumentally by human considerations of nature's 'usefulness' as means for human ends. Science strips the natural environment of beauty, power, and purpose and reveals nonhuman beings as 'mere entities' or raw materials. An early exponent of this view of Enlightenment science and technology is the German philosopher Martin Heidegger, who not only prefigured aspects of postmodern theory, but is seen as a central link in the relationship between postmodernism and environmental concerns

(Zimmerman, 1992, 1994). According to Heidegger, modern science and technology offer a misguided and dangerous view of the natural world, and the place and relation of humans in and to it.

At the same time postmodernism, again building upon critical theory, suggests that the domination of nature, which is at the heart of modernity, results in the domination of human beings. Thus the emancipatory aim of the Enlightenment, to free humanity from ignorance, fear, squalor and domination, actually creates a new system of ignorance, fear and domination. In this way, postmodernism challenges the idea of modernity being 'progressive' or better, and instead suggests that modernity and modern society are simply 'different' from other possible forms of socio-economic development and organisation.

Thus from a postmodern perspective the Western model of progress and development cannot be said to be 'better' or 'more advanced' than non-Western societies and non-Western models of development. This rejection of the simplistic equation of 'modern' (i.e. Western/Eurocentric) values, principles and socio-economic models with 'progress' is something postmodernism shares with radical green critiques of Western forms of development being imposed on the non-Western world. While the Enlightenment sought to universalise 'modern' values, principles and practices, in effect what this implied was the 'Westernisation of the world' (Latouche, 1993). This then lead to the creation of a hierarchy in which non-Western societies, values and practices were, by definition, 'non-progressive' and 'pre-modern'. In this way, for postmodernism 'modernity' and its allied concepts and practices of 'modernisation', 'progress' and 'development' are and have been used to create a situation in which the 'otherness' and 'difference' of non-Western societies leads them to be seen as unequal and inferior to Western societies and Western values and norms. Postmodernism's view is that 'difference' and 'otherness' (central valued categories within postmodern theory), demands respect and equality and is thus against the creation of a universal set of standards, values and practices which has the effect of homogenising both the human nonhuman worlds and erasing 'difference' and 'otherness'. A more recent instance of this has been the postmodern critique and rejection of **globalisation**, which from its perspective basically implies the Westernisation of the world and the universalisation of 'Western' model and standards of 'development' and what constitutes 'progress'.

The ecological sensitivity or potentiality of postmodern social theory is no where more evident than in its defence and valuation of otherness and difference. This is because the nonhuman natural environment is the ultimate 'other' for humans, since while our increased scientific knowledge of the natural world can mean we can learn more about it, ultimately we will never know the natural world fully, since a scientific understanding of the world is only one form of knowing. That is, from a postmodern perspective, the nonhuman world will (or ought) always remain, at least in essence, strange to us, mysterious and wonderful. We will never be able to possess or grasp it in its entirety. Thus while Donna Haraway, one of the leading postmodern theorists in this area, holds that, 'Nature is . . . that which we cannot not desire' (1995: 69), in keeping with postmodern theory, she also notes that a dominating attitude and treatment of the natural world can obscure as much as it reveals. Her view is that, 'We must find another relationship to nature besides reificiation, possession, appropriation, and nostalgia' (1995: 70), one based on a respect for the nonhuman world's very strangeness and alienness. However, basing how we ought to view and treat nature on a recognition of its alienness and otherness is not particular to postmodernism (see O'Neill, 1993).

## Postmodernism and post-industrialism

This idea that postmodernism represents a solution to the ecological crisis has expressed itself in many ways, particularly when postmodernism itself is related to 'post-industrialism'. Postmodernist social theory describes the present state of society, not only as 'postmodern' in cultural terms, but, related to that, also as post-industrial in terms of its economic system. Postmodernism in the words of Frederic Jameson (1991) is the 'cultural logic of late capitalism', where 'late capitalism' can be roughly equated with a 'post-industrial' stage of capitalism. The term 'post-industrial' conveys the idea that 'advanced' Western societies are now in qualitatively different developmental stage than the 'industrial' one, which roughly spans from the beginning of the industrial revolution to the 1980s. The post-industrial hypothesis is that as industrialised societies progress they move away from economies based on large-scale, factory-based, industrial manufacturing towards service-based, hi-tech forms of employment, production, distribution and consumption. One shorthand way of understanding post-industrialism is to say that it implies a shift in industrial society away from making things to providing services. Insofar as postmodernism implies post-industrialism, and insofar as green politics is a post-industrial politics with a vision of a post-industrial society, then to that extent postmodernism and green politics have much in common. The important point here is that 'post-industrialism' is typically argued to require a less exploitative use of the natural environment. A largely service-based economy, with modern electronic communications and information processing networks and technology, in which information, rather than 'raw materials' is the central economic resource, is argued to create a more sustainable economy.

The connection between postmodernism and green politics can also be seen in those analyses which suggest that the non-instrumental appreciation of the natural environment which is central to green politics, requires a certain degree of material affluence. In this way 'post-industrialism' is intimately related to 'post-materialism'. According to Ronald Inglehart's famous 'post-materialism' thesis, a profound shift took place in Western societies in the 1960s and 1970s, which helps explain the rise of green politics and a non-instrumental concern with the natural environment. Inglehart (1977), echoing some of the themes within Giddens's thought discussed in Chapter 4, suggested that the experience of peace, rising levels of material affluence, employment, mobility and education in Western societies since the Second World War, had led to a profound cultural change in those societies. Basically his thesis was that having secured high levels of material satisfaction and standard of living, many people in Western societies became aware of a lack of 'post-material', qualitative goods and experiences. Chief among these post-material interests was a concern with the preservation of the natural environment which led to a rise in green politics. In this way then, postmodernism is related to post-materialism and the consequent connection between the latter and ecological politics, interests and movements.

However, the equation of a concern with the environment and environmental politics as 'post-material' is something that can be criticised particularly when there are many environmental issues and movements which are characterised by 'material' as opposed to post-material interests and concerns. There is a very 'material' basis for environmental concerns and green politics, based on the relationship between environmental degradation, poverty, ill-health, and social injustice, aptly expressed as an 'environmentalism of the poor' by Guha and Martinez-Alier (1997).

## Postmodernism and the social construction of the environment

A final and arguably more contentious area of postmodern engagement with the environment involves its 'social constructionist' approach to environmental problems (Hannigan, 1995), and its deconstruction of such key terms as 'environment', 'nature' and the 'natural' (Robertson *et al.*, 1996; Andermatt Conley, 1997).

According to Hannigan, 'environmental problems are not very different from other social problems such as child abuse, homelessness, juvenile crime or Aids' (1995: 2), in that what we should focus on is not the 'problems' as such, but how the 'problems' come to be 'constructed' by the interaction of the social actors, the types of knowledge deployed and the power relations between these factors. An extreme version of the social constructionist perspective would be the postmodernist Jean Baudrillard's idea that environmental 'problems' are 'created' by images, the media and means of communicating/constructing them, not that there are objectively existing environmental problems 'out there' in nature for humans. While most social constructionist approaches, such as postmodernism, would not go so far as to deny the reality of environmental problems, the important point they raise is that in analysing environmental issues one must be aware of the different actors, claims, types of knowledge, communication and cultural contexts in which these problems are articulated, contested, presented and re-presented. The key to understanding the attitude of postmodernism to environmental issues is to remember that for it 'knowledge is power'. Hence from a postmodern perspective what is of primary importance is to identify whose or what knowledge is the dominant or most powerful in the social construction of the problem and suggestions for possible solutions.

At the same time, building on the critique of 'modernist' science in particular, and theories of knowledge in general, postmodern social theory has questioned the status and meaning of such central concepts such as 'nature', the 'natural' and 'environment' as well as 'human' and 'nonhuman'. In many respects the input of postmodern social theory to the analysis of environmental issues highlights the way in which these concepts have multiple meanings and senses. In this way, they are always open to alternative meanings and understandings. What 'nature' and 'environment' are and mean varies from culture to culture, group to group and means different things at different historical periods.

The multiple meanings of such terms as 'nature' and the 'environment' is by no means limited to postmodernism, indeed, the contested character of these terms was raised and discussed in Chapter 1. However, postmodernism has arguably gone further in exploring the multiplicity of meanings of concepts such as 'nature', 'environment' and the 'natural'. For example, what does 'nature' mean in an age when nature as an external, autonomous, independent entity is increasingly affected by human activity? What do 'natural' and 'unnatural' mean to different people? Why, in the Western world, is such a premium placed on all things 'natural' (goods, experiences, commodities, foodstuffs) precisely at a time when 'nature' is perceived to be under threat or disappearing?

## Problems of postmodern environmentalism

Postmodern 'deconstruction' and its view of the 'postmodern' social world as inherently fragmented, chaotic and unstable seems to fit with certain strands in modern scientific inquiry, such as chaos theory which also see the natural world as unpredictable, chaotic and unstable. However, some schools of postmodern thought which reject the independent 'reality' of the 'real world are deeply problematic from an ecological point of view which is premised on the non-discursive, non-negotiable existence of independent ecological, material limits on human action' (Gandy, 1996: 36).

As Zimmerman, in his study of postmodernism and radical ecology notes, 'radical ecologists criticize *some* aspects of modernity, while appropriating and transforming *other* elements of its emancipatory vision' (1994: 4). That is, critical social theory based on ecology and motivated by a concern for environmental destruction need not necessarily require the abandoning of the 'project of modernity'. Like Achilles' lance, modernity (as theorists such as Beck suggest with his theory of 'reflexive modernisation', and also the theories of Habermas and Giddens) is argued to be able to heal the ecological (and other) wounds it inflicts. Thus, there is no *necessary* connection between environmental concerns and politics and postmodernism.

The limits of postmodern analysis of environmental issues are no more clear than at the level of global environmental degradation. For some critics, postmodernism is unable to articulate a political economy of global environmental destruction, which leads to an inadequate analysis of power and social agency in a meaningful, useful and political sense. This 'apolitical' or 'anti-political' charge is a common one made against postmodernism. According to Gare, 'Postmodern culture is the culture of a society in which politics has become a farce, where rational critique and protest have become impossible' (1995: 34), and goes on to state that, 'Decrying the quest for political power as the problem, they [postmodernists] have handed over responsibility for their fate, and the fate of the world environment, to the economic rationalists, to the new international bourgeoisie and the international market' (1995: 34).

At the same time, focused as postmodernism is on the 'discursive' aspects of environmental politics and debates, leaves it unable to deal with the ecological, non-discursive, material dimensions of environmental problems (Gandy, 1996).

Postmodernism offers interesting and sometimes illuminating 'decon-structions' of how terms such as nature, environment, and related concepts such as human, are inherently *relational* in that one cannot speak of 'human' without also speaking of the 'nonhuman' or the 'natural' without its corresponding opposite of 'unnatural'. However, for many, its relativism, ambiguities, lack of a clear conception of political power and agency, its drift into a celebration of 'late capitalism' and its consumer culture, individualistic focus and stress on aesthetics and lifestyle means it is inadequate for the full range of issues around the relationship between human societies and their environments.

## Conclusion

This chapter has surveyed two recent social theories which analyse or can be used to analyse the environment and the relationship between environment and society. What both Beck's 'risk society' thesis and postmodernism share is that they are both orientated around the Enlightenment or modernity, and both explore how central to modernity and modernisation is the change that takes place in how society views, values and uses the nonhuman environment. However, while 'risk society' and the idea of 'reflexive modernisation' can be broadly seen as within the 'project of modernity', postmodernism, as the name suggests, is more critical of modernity, modern ways of thinking and acting in general, as well as in relation to the environment.

## Summary points

- A central point of reference in looking at the environment and social theory is modernity and its legacy.

- Beck's 'risk society' thesis both *describes* the current Western world as one in which people and politics are more concerned with the distribution of 'bads' (such as environmental and other risks) rather than 'goods' (income and jobs), and also *prescribes* a more ecologically enlightened social alternative.

- A central part of Beck's analysis focuses on how risk society calls for what he calls a process of 'reflexive modernisation' and the redefinition of what constitutes social 'progress'.

- Central to the latter is the necessity (as well as desirability) of spreading democratic control and accountability to more areas of decision-making which affect people's lives, particularly, scientific and technological developments.

- Whereas Beck's 'risk society' thesis argues that modernity and modern society based on its principles can solve its own environmental and other problems, postmodern social theory rejects the idea that modernity can do this, and instead calls for 'postmodern' solutions and approaches.

- Postmodernism is particularly good at indicating the multiple and contested meanings of such social constructions as 'environment', 'nature' and 'natural'.

- While postmodernism highlights the importance of identifying the discourses and powerful actors in environmental issues and disputes, it has an 'apolitical' or 'anti-political' character which is a serious flaw.

## Further reading

On Ulrich Beck's 'risk society' thesis and 'reflexive modernisation' see his *Risk Society: Towards a New Modernity* (London: Sage, 1992); *Ecological Enlightenment: Essays on the Politics of the Risk Society* (Princeton, NJ: Humanities Press International, 1995a); and *Ecological Politics in an Age of Risk* (Cambridge: Polity Press, 1995b). An excellent critical analysis of Beck can be found in ch. 5 of David Goldblatt's *Environment and Social Theory* (Cambridge: Polity Press, 1996). Other sources include: Scott Lash, Bron Szersynski and Brian Wynne (eds), *Risk, Environment and Modernity: Towards a New Ecology* (London: Sage, 1996).

On postmodernism and environmental issues see: Arran Gare, *Postmodernism and the Environmental Crisis* (London: Routledge, 1995); Michael Zimmerman, *Contesting Earth's Future: Radical Ecology and Postmodernity* (Berkeley: University of California Press, 1994); Klaus Eder, *The Social Construction of Nature: A Sociology of Ecological Enlightenment* (London: Sage, 1996). The work of Donna Haraway is particularly good as an example of how aspects of postmodern social theory can be used to analyse the environment, and social-environmental relations. See her *Simians, Cyborgs and Women: The Reinvention of Nature* (London: Free Association Books, 1991).

Other postmodern approaches to theorising the environment include: Vera Andermatt Conley, *Ecopolitics: The Environment in Poststructuralist Thought* (London: Routledge, 1997); G. Robertson *et al.* (eds), *FutureNatural: Nature, Science, Culture* (London: Routledge, 1996); and John Hannigan, *Environmental Sociology: A Social Constructionist Perspective* (London: Routledge, 1994).

# 8 Ecology, biology and social theory

- Sociobiology: genetics, 'human nature' and ecology
- Sociobiology and the biophilia hypothesis
- The appeal of human nature in social theory
- Human embodiedness and embeddedness: from a dualistic to a dialectical culture/nature relation
- Transforming nature: the creation of our ecological niche
- Human flourishing and the natural environment
- Human categorisation of the environment: resources, non-resources and proscribed resources
- Naturalism and social theory

## Introduction

Having introduced some of the ways in which the environment has been conceptualised with some central strands of twentieth-century social thought, this chapter continues this exploration, but widens out the issue beyond the 'environment' (both natural and urban), to look at approaches to the study of the natural world and humans in relation to and a part of that world, and how natural sciences (such as biology, ecology and ethology) have influenced social theorising and/or been used themselves as the basis for social theory.

For many social theorists, the social-scientific study of the relationship between society and the natural environment requires that we address the following question: 'how do we open up to investigation the relationships between humans and the rest of nature, without letting in the "Trojan horse" of biological determinism?' (Benton and Redclift, 1994: 4). That is, how are we to study the ecological relationship of the human species to its natural environment (or ecological niche) while avoiding thinking about and studying humans as if they were simply another species and thus explain human behaviour using the same biological and ecological models and conceptual analysis used in the case of other species? At the same time it is quite obvious that humans are a particular species, living in particular environments like other species on the planet, and thus it seems odd to reject a biological or ecological approach to the study of human society and behaviour. So why are we worried about 'biological determinism', if humans are a biological species, and the behaviour of other biological species can be explained and theorised on the basis of their biological characteristics within the context of their environment? But are humans the same as other species? Surely, some would say, humans are 'special', 'different' in some important way from other species, which makes the application of biological and ecological models inappropriate to them? This chapter seeks to discuss these questions and examine the various ways recent theorists have sought to integrate biological and ecological insights into the study of human behaviour and relations between human society and the environment.

## Sociobiology: genetics, 'human nature' and ecology

While Benton and Redclift (1994) are weary of adopting a 'biological determinist' approach in studying human-environment relations, for some this question is unproblematic. For some a 'biological determinist' approach is absolutely central to understanding not only human relations to our environment, but also to explaining our social relations to fellow members of our species. The school of thought which best expresses this 'biological determinist' view is sociobiology, which as its name suggests is a composite theory of human behaviour in which biological theories are applied to the human social world. As the *Hutchinson Dictionary of Ideas* puts it, sociobiology is the 'study of the biological basis of all social behaviour, including the application of population genetics to the evolution of behaviour' (1994: 483). For sociobiology, to deny our biological make-up, our 'selfish genes' in Richard Dawkins's (in)famous term, in our analysis of human society (and especially in our proposals for social change) is a dangerous mistake.

Sociobiology as a form of 'biological determinism', basing its claims on the direct application of genetic and evolutionary theory to humans, has a long pedigree, and has clear antecedents in Malthus's 'theory of population', 'Social Darwinism' and Herbert Spencer's ideas, as discussed in Chapter 3. A flavour of its biological-ecological determinism can be seen in Wilson's claim that, 'it is entirely possible for all known components of the mind, including will, to have a neurophysiological basis subject to genetic evolution by natural selection' (1978: 10). Taking human beings and their social and individual behaviour as just like the behaviour of other animals, sociobiology has sought to explain society, politics, culture, and the relationships between men and women in terms of genetics and evolution by natural selection.

In short, sociobiology advances a particular idea of 'human nature' and uses this to explain human social life. As a form of social theory, sociobiology aims to explain social phenomena by insisting that humans are not just *like* other animals and therefore analogies can be drawn between humans and comparable species, but that we *are* a species of animal, with our own natural characteristics, and to deny that animal nature is mistaken. As Gregory explains, 'sociobiology tells us much about the inherited behavior of animals; and it tells us in a new way that we ourselves are animals' (1978: 293–4). Its view of human nature as a 'fixed' essence and largely immune to the effects of socialisation or 'nurture' shares many of the aspects of the traditional conservative political view in which a similar 'negative' view of human nature is used to criticise political programmes which play great faith in the 'innate goodness' of humans and/or the capacity of altered social practices and institutions to make people less selfish, violent, competitive and so on. In this respect, sociobiology often seems to echo Malthus's in his critique of 'progressive' Enlightenment thinkers of his time. Like Malthus, sociobiologists speak of natural constraints (especially the evolutionary, genetic make-up of humans) which they argue undermines attempts to make the world a better place through changing socialisation patterns or the social environment of individuals. Whereas for many socialists, people can be encouraged to behave in a more cooperative manner by changing their social environment, for sociobiologists what's more important is the evolutionary and genetic aspects of human behaviour (which in part is based on the 'original', prehistoric natural and social environment of the evolving human species).

For Dawkins, another early proponent of sociobiology, human behaviour is explicable in terms of the 'selfish gene'. Dawkins's claim is that human individuals are simply carriers of genetic information, and since the principal aim of genes is to reproduce themselves, all of human behaviour, social rules, taboos, traditions, institutions, etc., can be explained by reference to the 'selfish gene'. Biology determines society, and in this way culture is reduced to human nature. One of the interpretations of the modern ecological crisis, from a sociobiological perspective, is that our genetic coding, or instincts evolved in the 'hunter-gatherer' stage of human evolution, are not suited or do not exhibit 'inclusive fitness' in the modern situation. Our advancement, technologically, scientifically, economically and socially, in radically altering our relationship to the natural environment, has 'outstripped' our instinctual ability to cope with these changes (Berry, 1986).

The idea of human nature within sociobiology comes down to the 'instinctual' basis of human action, that these are somehow 'given' and therefore 'beyond change'. The import of this as Berry has noted, is that 'Humans as humans have a limited range of options open to them. The presence of this limited range means that there is a constancy and predictability about what humans will do in specific situations' (1986: 104). Often sociobiological arguments express themselves not so much in the idiom of 'biological' as 'psychological determinism'. According to Davies (1976), the Russian revolution (or at least Lenin's part in instigating it) can be explained along the following lines, 'As a very young man, Lenin experienced the death of his father, his brother, and his sister, for reasons attributable to the Tsarist government. Decades later, these experiences became part of the fuel of a revolution' (1976: 104). Such controversial social theorising is part and parcel of sociobiology, controversial in reducing social events, such as revolutions, to matters of individual psychology and personal experience. The point is not that the latter are not important in explaining social action, but rather one cannot *reduce* the latter to the former.

For some sociobiologists, the 'fact' that humans (or at least males) are 'naturally aggressive' and 'competitive' is something that is fixed and 'given' and therefore unchangeable. As these behavioural traits are genetically 'hard-wired', the best society can do is either to minimise their negative impacts or use them to produce/motivate socially desirable outcomes. Echoing aspects of Freud's work, sociobiology suggests that since war and aggression is 'natural to humans' and instinctual, the best we can do is sublate or channel these potentially harmful impulses into socially beneficial or harmless pursuits. It from this type of argument that we come to the popular sociobiological explanation for competitive sports in human cultures as evolutionary adaptations to innate human aggression, group loyalty, group and individual competitiveness and so on. Football, on this view can be explained as a cultural phenomenon in which natural (male) impulses for male bonding, competitive and aggressive territorial behaviour, can be exercised and (neutralised) on (but increasingly spills over and off) the pitch.

At the same time, sociobiology can also give an evolutionary-genetic reason for the predominance of males in warfare. While the orthodox sociobiological explanation for this is on the basis of the superior physical strength of (young) men as compared with women – that is men are 'adapted' for warfare – there is another less 'sexist' interpretation. On this view (partly based on anthropological evidence), the reason why young men have traditionally been the 'warriors', and have monopolised (and still do) warfare and the military institutions of society (that is, the institutions of organised violence and aggression) is that unlike women, young men are 'superfluous' from the genetic and evolutionary perspective. That is, sending a large proportion of young men off to war (with a good chance of them being killed) did not significantly affect the genetic evolution of the species. At the same time, anthropological evidence seems to suggest that in 'hunter-gatherer' societies (the societal level in which humans have spent most of their evolution), it was women who provided most of the protein in these societies (from gathering vegetables and rudimentary cultivation) and not the occasional and infrequent contributions of meat from male hunting (Diamond, 1991: 34).

Many have criticised some sociobiological arguments as based on ideological rather than scientific theory. Sociobiology is held to offer 'scientific' legitimacy to right-wing, conservative and 'free market' ideology (Alper, 1978; Berry, 1986: 109). One can see here in the language of the 'selfish gene', the innate competitiveness and aggression of human beings, an echo of the 'invisible hand' idea of the eighteenth century, where the self-interest of each person will, in market transactions, result in socially beneficial results. F.A. Hayek, one of the foremost theorists and defenders of a libertarian, free market perspective represents a good example of this ideological appropriation. As Dickens (1992) points out, Hayek's evolutionary view of human beings leads him to see the modern capitalist system, the free market and private property as adaptive forms of what he calls 'the extended order', a cultural form in which civilisation, social order human progress and development have been achieved. Hayek's 'organic' view of human society, in which society is seen as a 'super-organism' in relation to its environment, places this strand of right-wing thought within strands of conservatism, which also have an 'organic' view of society. This 'naturalistic' view of society is used to combat those (usually from the left of the political spectrum) who argue that conscious collective action, altered social institutions and practices can create a new type of society and human being. For the conservative, such 'social engineering' is both impossible and dangerous, since the essence of human beings, their natures, are fixed and immune from deliberate social changes.

Sociobiological arguments have caused great controversy, particularly when used in arguments about whether 'nature' or 'nurture' determines human behaviour. The political importance of this debate should be obvious. If 'human nature' is fixed genetically and behaviour reduced to genetic causes, then there is little point in trying to create a 'social environment' aimed at encouraging people to be less selfish, competitive and so on. Trying to make people less selfish is seen by those who subscribe to the sociobiological/conservative view as 'unrealistic' and 'utopian', in attempting to 'go against the grain' of human nature.

A particularly contentious example of sociobiological theorising is its use to scientifically 'prove' innate, 'natural' differences between different racial or ethnic groups. Particularly in America there have been attempts to prove that Afro-Americans are genetically less intelligent than whites or Asians. This is the basic argument of a book by Charles Murray, *The Bell Curve*, which was criticised (like much crude, deterministic versions of sociobiology) for being right-wing and reactionary, and for justifying racism and racist attitudes and finally for misusing science in order to do so.

## Sociobiology and the biophilia hypothesis

There have been attempts, notably by E.O. Wilson, to establish a genetic or evolutionary reason/explanation for what he calls 'biophilia' or love of nonhuman life. Now while there are many genuine questions one can ask concerning the evidence for this genetic, hard-wired concern for nonhuman life, whether Wilson is talking about 'biophilia' or mere 'bioconnection', and whether this genetic 'ecological sensitivity' is relevant to modern, urban dwellers (see Baxter, 1999: ch. 3), it still remains the case that biophilia is a sociobiological attempt to explain why we do have a (genetic) concern for the nonhuman world. According to Wilson,

> *Biophilia, if it exists, and I believe it exists, is the innately emotional affiliation of human beings to other living organisms.* Innate means

> hereditary and hence part of ultimate human nature. Biophilia, like other patterns of complex behavior, is likely to be mediated by rules of prepared and counter-prepared teaming – the tendency to learn or to resist teaming certain responses as opposed to others. From the scant evidence concerning its nature, biophilia is not a single instinct but a complex of learning rules that can be teased apart and analyzed individually. The feelings moulded by the learning rules fall along several emotional spectra: from attraction to aversion, from awe to indifference, from peacefulness to fear-driven anxiety.
>
> (emphasis added)[1]

In other words, biophilia is a love of, or concern with, the nonhuman world which is rooted in human nature. At root, the genetic explanation for something like biophilia is, as Baxter (1999: ch. 3) explains,

> human groups whose individual members possess something like an epigenetic reverence for nonhuman nature, at least to a degree, may produce better than groups whose members possess a purely instrumental attitude towards it . . . Evidence also compatible with it may be drawn from the fact that even in our modern artifactual world, where individuals may live their entire existence in an almost artificial environment, people show a yen for, and exhibit moral attitudes towards, the nonhuman natural world. They make gardens, keep house plants and domestic pets, go for country walks, watch natural-history films and so on.

According to Wilson there are evolutionary grounds for establishing 'kinship' with other nonhuman species.

> Other species are our kin. This perception is literally true in evolutionary time. All higher eukaryotic organisms, from flowering plants to insects and humanity itself, are thought to have descended from a single ancestral population that lived about 1.8 billion years ago . . . All this distant kinship is stamped by a common genetic code and elementary features of cell structure. Humanity did not soft-land into the teeming biosphere like an alien from another planet. *We arose from other organisms already here, whose great diversity, conducting experiment upon experiment in the production of new life-forms, eventually hit upon the human species.*
>
> (emphasis added)[2]

This appeal to there being 'something' in human nature (its genetic and evolutionary history) which means that having a relationship to the natural environment is important is something shared with other 'naturalistic' approaches to social theory discussed below.

For Wilson, biophilia can be proved to have a genetic basis. For example, he claims that fear, and even full-blown phobias, of snakes and spiders (but also other phobic elements of the natural environment such as dogs, closed spaces, running water, and heights) are quick to develop with very little negative reinforcement, while more threatening and risky modern artefacts, such as knives, guns, electric wires, and cars, do not usually evoke such a response. Humans find trees that are climbable and have a broad, umbrella-like canopy (such as oaks, poplars) more attractive than trees without these characteristics (such as firs and pines). Finally, people would rather look at water, green vegetation, or flowers than built structures of glass and concrete.

For Wilson (and an ecologically orientated sociobiolgy), the biophilia hypothesis, if substantiated, provides a powerful argument for the conservation of biological diversity, on the anthropocentric (human-centred) grounds that the maintenance of a diverse and richly varied biological range is genetically important for human (psychological, emotional and perhaps spiritual) well-being, as well as being important for providing the ecological resources for society.

## The appeal of human nature in social theory

The appeal of using 'human nature' in social theory is understandable. Human nature is universal, that is, it is something shared by all human beings regardless of culture. Thus any assertion or claim based on an appeal to human nature is universalisable to humanity as a whole and not just to one particular society in a particular historical period. Hence the political importance of 'human nature' is that it is at *one and the same time a descriptive and a prescriptive concept.* That is, when one says, for example, that it 'goes against human nature' for women to work outside the home, have careers rather than raise children, one is both describing (explaining) the gendered division of labour between the sexes, but also saying this particular division ought to be. As I am sure most readers will recognise, in popular consciousness and popular argumentation there is often an appeal to 'human nature' either to explain or justify some particular position or to refute competing arguments. Indeed, the appeal to human nature is perhaps the most ubiquitous form of justification and or explanation within debates about a variety of social phenomena.

As Berry notes, 'the notion of human nature, with its assertion of the givenness of man, which serves to demarcate the boundary between the supposedly remediable human world and the supposedly irremediable natural world, is essentially an ideological construct' (1986: 105). The problem with sociobiology as a form of social theory is that it attempts to explain everything about human social life in terms of genetic and evolutionary factors. As Gregory explains, 'One of the problems sociobiology encounters in seeking genetic determinants of behavior is that it must explain everything or else it explains nothing' (1978: 286). Hence sociobiology's attempt to explain everything from human warfare, the social division of labour, to individual psychological problems in terms of genetics, 'adaptive fitness' and so on. And across a wide variety of social phenomena, its explanatory powers have, to say the least, been found wanting (Dickens, 1992; Rose *et al.*, 1984). It is not that sociobiology is necessarily wrong in wishing to break the barrier between the 'social' and the 'natural', but rather how it simply *reduces* the former to the latter on the basis of a very deterministic account of human nature.

However, to reject the biological or genetic determinism of sociobiology does not mean that we need to deny that human beings are 'natural' or indeed that 'culture' and human cultural practices are somehow not natural. Sociobiology in some senses has to be recognised as representing one of the main attempts to transcend the 'nature/culture' dualism, though misguided in simply reducing culture to nature in general and evolutionary genetics in particular.

## Human embodiedness and embeddedness: from a dualistic to a dialectical culture/nature relation

In this section, we look at an alternative view of the relationship between social theory, biology, ecology and the environment. Over the last twenty years or so Ted Benton has sought to integrate biological and ecological ideas into social theory, from a broadly left-wing perspective. His starting point is one which recognises the perils of what he calls an over-naturalistic perspective in which, 'the concepts of ecology as a biological science apply in an unqualified way to the human species, whilst the philosophical principles underlying ecology are generalizable as a set of norms for human conduct' (1994: 39). He rejects Malthus's population theory (and sociobiology, discussed below) for applying unmodified the concepts of natural science to the study of human society. At the same time, he does not reject the importance of biological and ecological considerations. As he puts it,

We can, and, I think, we should, continue to view humans as a species of living organism, comparable in many important respects with other social species, as bound together with those other species and their bio-physical conditions of existence in immensely complex webs of interdependence, and as united, also, by a common coevolutionary ancestry. To say this much *is* to be committed to a naturalistic approach, but not necessarily to a reductionist one. It is to be committed to recognizing the *relevance* of evolutionary theory, physiology, genetics and, especially ecology itself, as disciplines whose insights and findings are pertinent to our understanding of ourselves.

(Benton, 1994: 40; emphasis in original)

Thus Benton argues that we should not confuse the relevance of the natural sciences in investigating social phenomena in the human world with these sciences offering a full explanation of those human phenomena. This is the mistaken path taken by sociobiology. Though sharing with nonhuman species similar problems and similar characteristics, biological needs for food, for example, humans are *not* the same as other species. Our particular natures, needs and modes of flourishing are such that we are different from other species. But being different does not mean that we as a species are somehow radically separate or 'superior' to other species. In this way, Benton, like others in this chapter, seeks to transcend the dichotomy between 'environment' and 'society', as an important step in developing a more ecologically sensitive form of social theory. Yet as many feminists have noted 'biology is not destiny' and neither is ecology for that matter, and Benton is aware that one equally must recognise that there is both a *difference* and a *connection or commonality* between the social and natural worlds. One way of looking at this is to see that *human beings are a part of, as well as apart from the environment* (Barry, 1995c).

This latter injunction, to see humans as at one and the same time a part of the natural order (and hence like other species) as well as being apart from it (and hence not being like other species), has also been taken up by others such as Hayward (1995), Dickens (1992) and Brennan (1988). Brennan neatly sums up one of the main reasons for adopting a naturalistic approach: 'in order to discover what sort of human life is valuable we must first consider what kind of a thing a human being is. Although there is, in my view, no complete answer to this question, we can . . . grasp one important aspect of human nature by reflecting on what are essentially ecological considerations' (1988: xii).

An example of this position, and one which illustrates how Benton and others have sought to integrate biological and ecological considerations into social theory (as opposed to reducing social theory to these natural sciences), is the human requirement for food as a condition for survival and flourishing. While, like other species humans require food, in the case of humans it is not food per se that we need, but food collected, prepared and consumed in particular ways. As Benton puts it, 'Proper human feeding-activity is symbolically, culturally mediated' (1993: 50). The importance of this point is its sensitivity to ecological and biological conditions for human need-fulfilment. A graphic illustration of this is the title of Lévi-Strauss's famous book *The Raw and the Cooked*, the 'raw' referring to the mode of food consumption within the nonhuman world, and the 'cooked' to the mode of consumption within human societies.

On this view, culture can be regarded as our species-specific mode of expressing our nature, or 'species-being'. As it is continuous with our nature as social beings, human culture does not represent a radical separation from nature, but rather can be viewed as our 'second nature' (Bookchin, 1986), that is, emerging from, but situated within, the natural order. The importance of this has been expressed by Kohák who notes that, 'Were culture a negation of nature, no integration of humans and nature would follow' (1984: 90). Benton seeks to overcome the 'nature/environment' versus 'human/culture' divide by articulating a naturalistic account of human beings in which human needs, conditions of flourishing and human culture can be understood 'naturalistically'. Benton's naturalistic social theory begins from his observation that, 'Humans are necessarily *embodied* and also doubly, ecologically and socially *embedded*, and these aspects of their being are indissolubly bound up with their sense of self and with their capacity for the pursuit of the good for themselves' (1993: 103; emphasis in original). In drawing attention to the 'embodiedness' of human beings, Benton is close to the materialist ecofeminist arguments outlined in Chapter 5, in which human embodiedness is used to highlight and criticise the 'sexist' assumptions of social theories which deny the neediness, vulnerability and dependency of human beings.

Benton criticises what can be called the foundational separation between 'culture' and 'nature' that is the basis of social science (as distinguished from natural science) in general, and social theory in particular. The clearest example of this separation which Benton criticises is that between humans (their unique capacities, needs and forms of sociality, etc.) and animals. While not for one moment denying the uniqueness of the human species, Benton does point out that across a broad range, the specific capacities and qualities often assumed to be unique to human beings can also be found in other species. So, for example, he states that 'The capacity for and disposition to social co-ordination of activity as such is not a distinctive feature of our species' (1993: 36), since many other social species, such as primates, also exhibit often highly complex forms of social coordination. Challenging one of the central claims of contemporary social theory, which not only separates human society from nature, but also denotes human capacities and behaviour as 'higher' or superior to that associated with nonhumans, Benton suggests that 'those things which only humans can do are generally to be understood as rooted in the specifically human ways of doing things which other animals also do' (1993: 48). In this way, as pointed out above, like other forms of organic life humans need to eat food, but unlike these other species we have a variety of ways in which food is 'cooked' (culturally mediated) rather than consumed 'raw' (non-culturally mediated).

Another example, of the needs and activities of human beings as 'differentiations' rather than unbridgeable 'differences' in comparison with nonhuman species is reproduction and sex. According to Benton, 'continuous sexual receptivity [suggests] an evolutionary significant role for sexual activity in human social life to some extent dissociated from the immediate requirements of reproduction' (1993: 53). Thus unlike other species, in which the female of the species is sexually receptive for limited periods, and in which sexual activity is orientated towards reproduction, humans have a much more complex orientation towards sexual behaviour. The complexity of sexual activity within human society, can in part be taken as indicating that sex for humans is not simply driven or motivated by reproductive concerns (an instrumental orientation to sex), but has a deeper evolutionary significance in that it is invested with cultural or social meanings. Sex in human societies is governed by normative and socially authoritative sanctions, rules, taboos; hence the existence of an incest taboo in almost all human cultures and the 'private' character of sex within human societies. And one can think of many other ways in which sex is *meaningful* for humans in ways which it is not for other species, where it is the *means* by which reproduction is achieved. However, in this case, one could suggest that the difference between humans and nonhuman animals is a *difference in kind rather than degree*. However, this does not undermine the naturalistic basis of Benton's approach, in which the important point is that sex for the human species is not, from an evolutionary and naturalistic perspective, solely (or perhaps even primarily) explicable in terms of reproduction. On the other hand, the *centrality* of sex (and its various manifestations of sexuality and sexual activity) within human cultures *is* something that is shared with other species where sex (qua reproduction) is a central organising need and activity (even if only infrequently, but regularly practised).

## Transforming nature: the creation of our ecological niche

From a naturalistic perspective, human culture can be seen as a collective capacity of the human species to adapt to the particular and contingent conditions of their collective existence, including, most importantly, the natural environments with which they interact and upon which they depend. Thus, culture is in part the particular mode by which humans adapt to their 'ecological niche', but not simply in the sense that cultures are somehow 'determined' by environments. Rather, in the additional sense that the mode of human adaptation to their 'ecological niche', and the expression of their 'species being', involves the *active transformation of their environment*. As the biologist Lewontin puts it,

> We cannot regard evolution as the 'solution' by species of some predetermined environmental 'problems' because it is the life activities of the species themselves that determine both the problems and the solutions simultaneously . . . Organisms within their individual lifetimes and in the course of their evolution as a species do not *adapt* to environments: they *construct* them. They are not simply *objects* of the laws of nature, altering themselves to the inevitable, but active *subjects*, transforming nature according to its laws.
>
> (Quoted in Harvey, 1993: 28)

The dialectical character of the relationship and interaction between society and environment can be seen to be consistent with the 'new biology' which unlike earlier biological theory and sociobiology does not take there to be two separate entities to be observed: the 'organism' and the 'environment'. As Lewontin puts it, 'it is impossible to describe an environment without reference to organisms that interact with it and define it. Organism and environment are dialectically related. There is no organism without an environment, but there is no environment without an organism' (1982: 160). The point here is that there is no 'raw nature' to which organisms adapt. Rather the relationship is reciprocal or dialectical. Organisms change and make their 'natural environment', and at the same time the natural environment influences their behaviour and sets limits and opportunities on what organisms can do.

This dialectical relationship between organism and environment helps us to understand why there is no one 'ecological niche' for humans, as can be readily seen in the success of our species's colonisation of even the most inhospitable parts of the earth's surface. As a species nature did not specialise; our ability to transform the environment means that we are unique in the range of ecological niches in which we can flourish, present ecological problems notwithstanding. That is to say, the 'environment' for humans is not some 'raw' or untransformed one as given by nature, but rather a humanised one. Like other species, humans are not simply faced with the problem of how to 'fit' a particular 'given' environment. The environment is not 'given', fixed and something external to 'society', but nor is it completely 'materially (and socially) constructed', that humans can ignore and/or make for themselves. A dialectical understanding of the relationship between society and environment sees both as interdependent, the environment, in part, transformed (as opposed 'made' from scratch) by human activity, and the environment in turn providing opportunities and constraints for human activity.

## Human flourishing and the natural environment

Another aspect of this naturalistic social theory is Benton's idea that human flourishing has an ecological basis (but not simply in the sense that the natural environment transformed and worked upon by humans provides food, clothing, shelter and so on). At the same time, humans also need to relate to their environment culturally and symbolically as a condition for human flourishing. According to Benton, 'our interaction with and symbolic investment in our external environment are essential to the formation and maintenance of a stable personal identity' (1993: 181). From this perspective, alienation or estrangement from the natural environment can lead to a crisis of personal identity, mental breakdown and psychological distress. This point has also been made by environmental psychologists such as Kidner (1994) and others such as Goodin (1992: 37–41) who stress the 'human need' to find meaning by conceiving of themselves as part of a 'larger order of things'. The argument is that one's 'sense of self' is in part tied up with one's 'sense of place'. In this way, human personal identity and psychological well-being have an ecological basis. Thus, for Benton the natural environment becomes an essential part of a naturalistic understanding of human well-being and a condition for human flourishing. As he puts it,

> These features of the physical and social world which enter into and
> constitute our sense of self are not dispensable features which we may not
> choose to value or assign significance to. They are features which are
> basic in the sense that only in virtue of their presence can we hold on to a
> sense of ourselves as choosers, valuers and assigners of significance at all.
> (Benton, 1993: 184)

Hence the contributory factor of alienation from the natural environment
in the high proportion of mental (and other forms of) illness and
psychological distress in heavily urbanised areas. This is particularly the
case in certain forms of land use and urban developments such as high-
rise tower blocks and suburban housing estates. The importance of access
to the natural environment for human psychological health can also be
demonstrated negatively by reference to the individual experience of
nature as a central therapeutic stage in the treatment of some forms of
psychological distress and illness. It is important to notice here that these
same sources of ecologically based psychological harm are particular to
humans (though perhaps shared to some degree with apes and
chimpanzees). Unlike other species, humans do not simply need an
'ecological niche' or 'habitat', but over and above that humans require a
'sense of place' as a condition for their flourishing. To put it another way,
situating humans within a suitable ecological niche (though this niche is
not 'given' by the natural environment), is a necessary but not a
sufficient condition for their flourishing. Humans may survive within
such a natural environment, but without a sense of place they will not
flourish. Part of this distinction between 'survival' and 'flourishing'
(which of course is not limited to the human species), rests on the ways
in which our species relates to the natural environment in cognitive or
symbolic-cultural ways. Such cognitive appropriations of nature include
the scientific and aesthetic appreciation and experience of the natural
world, and other cultural modes of apprehension, valuing and
experiencing the natural environment.

At the same time, from the perspective of the late twentieth century, the
natural environments faced by human societies cannot be taken to be
completely and purely 'natural', i.e. independent of human influence. In
most cases, they have been transformed by past and current human
behaviour. There is a real, material basis to the claim that the
environment is 'socially constructed'. It is not just that our understanding
of the environment is mediated by human social relations and culturally
symbolic meanings, but the environment faced by human culture is often
partly the 'product' of previous social modification. It is frequently

difficult to maintain a strict division between a 'natural' environment and one which is the outcome of human purposive action in conjunction with that natural or given environment. The ecological niche for humans is as much a 'humanised' as a 'natural' one, as Lewontin suggests above. A naturalistic social theory views this transformative activity as central to understanding human nature and human culture. The significance of this cultural dimension is that the human 'ecological niche' is *both* culturally *and* biologically-ecologically determined. While like other species we must use the natural environment in order to live and to flourish, we do not react uniformly to it, and we are unique in our species's adaptability to different ecological conditions, and above all in our species's ability to transform the natural environment to our particular needs. As Woodgate and Redclift put it,

> we are uniquely equipped to regulate and refashion the environment in ways that make it more suited to our requirements. *Thus, there is no single way in which we, as human beings, relate to external nature.* Acceptance of the complex, interactive character of social and environmental change, means that simple distinctions between 'social' and 'natural' sometimes become untenable.
>
> (1998: 8; emphasis added)

This echoes Benton's suggestion that,

> *there is no 'natural' mode of human relation to nature.* No original, ecologically 'harmonious' golden age or state of grace from which we have fallen. Humans have no single, instinctually prescribed mode of life, but a range of indefinitely variable 'material cultures' . . . The forms of human ecology, as culturally mediated relations to physical, chemical and biological conditions, are both limitlessly variable *and* ecologically bounded.
>
> (Benton, 1994: 43; first emphasis added)

There is no single way in which humans relate to the natural environment as even a cursory examination of the variety (though this variety is not unlimited) of past and present modes of human interaction with the environment will illustrate. Thus there is no 'species-specific' or 'naturally given' manner in which the human species *does and ought* to interact, both materially and symbolically with the natural environment. The importance of this is that, unlike other species, one cannot specify what the 'proper' mode of human interaction with the natural is or ought to be. The mistake of reductionist accounts of human social behaviour such as sociobiology, but going back to Malthus and Spencer, is that the relationship between other species and their environments cannot be used to 'read off' or 'determine' the relationship between human society and its environment.

Population-environment relations in terms of the 'carrying capacity' of the natural environment to support some level of population is different in the human and nonhuman cases. As Benton points out, 'Human inventiveness, with respect to our powers of intentional modification of our environments through normatively ordered social practices, renders quite illegitimate any attempt to read off from a specification of the bio-physical environment what its "carrying capacity" might be for human populations' (1994: 42–3). Benton's reticence about using the carrying-capacity idea might also have something to do with the fact that its origins are ideological rather than scientific. According to Bandarage,

> The concept of 'carrying capacity', for instance, was first put into use by French and British colonial scientists and administrators seeking to estimate the minimum amount of land and labor needed by local people to meet their subsistence needs so that what was deemed in excess of that could be taxed by the colonial state and appropriated for export production.
>
> (1996: 127–8)

However, Benton's point is that in order to find out the 'carrying-capacity' of a particular environment for humans one must find out about the social relations, moral codes, cultural values and practices, economic arrangements, agricultural practices, property relations, forms of scientific and technological capacities and so on. Thus while a particular environment may support a particular human population, under changed social, economic, political and cultural conditions, this same environment may support a higher or lower population level. Our 'second nature' (to use Murray Bookchin's (1986) term for human culture/society) modifies or transforms the opportunities and constraints that 'first nature' presents to us.

At the same time, this 'indeterminacy' and variety it gives rise to in the way in which humans use and interact with the environment is in part due to the fact that we not only transform the environment, i.e. materially interact with it. Human relations with the natural world are also, in part, regulated by 'symbolic' interaction with and conceptions of the natural environment. We are unique in having a variety of 'cultural' and 'moral' norms, rules, taboos and so on which influence our material transformation and use of the natural environment.

# Human categorisation of the environment: resources, non-resources and proscribed resources

Though an oversimplification, generally speaking human beings and societies divide 'the environment' into three broad categories, 'resources', 'non-resources' and 'proscribed resources'. Other species divide their environment into the first two, basically that which they can eat and/or use in some way, and that which they cannot eat and/or use. Humans on the other hand, while fulfilling their needs (many of which they share with other species, such as the need for food, safety and reproduction) by using the environment do adopt a discriminating attitude to the natural world on the basis of 'non-instrumental', cultural or symbolic grounds in removing parts of the environment which could be used as 'resources' into the category of 'proscribed resources'. So, for example, we find human cultures in which some animals, which are otherwise perfectly suitable as sources of protein or hides, are not considered as such, as in Hindu cultures where the cow is seen as 'sacred' and thus placed in the 'proscribed resources' category. Other examples include the ubiquity within human cultures of denoting some nonhuman species as 'vermin', a cultural category which only makes sense against a background of there being 'non-vermin' species and 'non-vermin' modes of treatment. According to Diamond, 'the notion of vermin makes sense against the background of the idea of animals in general as not mere things. Certain groups of animals are then signalled out as *not* to be treated fully as the rest are, where the idea might be that the rest are to be hunted only fairly and not meanly poisoned' (1978: 476; emphasis in original). The important point is to recognise that the human species is set apart from other species by the ubiquity of this category. That is, in all human cultures we find the category of 'proscribed resources'. Figure 8.1 gives a brief overview of this simple model outlining the different categories which are important in the metabolism between species and the environment.

The usefulness of this model can be used to outline the main issue at stake in current debates about biotechnology and genetic engineering. In terms of the tripartite division of the natural world, the technological advances represented by biotechnology can be seen raising problems and opportunities which cut across all three categories. For those in the biotechnology industry one could say that what they are doing is moving genetic information which was previously in the category of 'non-resource' (since we did not have the technological capacity to use genetic

|  | Rats | Humans |
|---|---|---|
| **Resources** | Most organic matter, including waste and inorganic nest-building materials | Organic and inorganic material and energy |
| **Non-resources** | Materials not biologically-suited (e.g. inorganic matter) | Not 'given', i.e. dependent upon human biology (depends on technological factors) |
| **Proscribed Resources** | Not applicable | Cultural/moral variations (e.g. Hindus and cows, vegetarians, vegans). Genetic information as standing between 'resource' and 'proscribed resource' |

**Figure 8.1** *Species Metabolism with the Environment*

*Source:* Author

information) into the category of 'resource'. At the same time, many of those who raise moral or prudential reasons against genetic engineering, can be either seen to wish genetic information to be placed in the 'proscribed resources' category (that is, we have the technological potential to use this information as a resource, but choose not to) or advocate the abandoning or scaling down of biotechnology and genetic engineering, thus maintaining genetic information as a 'non-resource'. In this way, this discriminating attitude humans adopt towards their environment (an environment that is often transformed to suit their own particular purposes) illustrates the importance of the symbolic, cultural or moral dimension to the social-environmental metabolism.

The simple tripartite model in reference to the inclusion of a cultural or moral aspect to social-environmental relations, also illustrates another weakness of sociobiological arguments which depend on 'hard-wired' or an 'instinctual' basis for human behaviour. According to Benton,

> Lacking what Mary Midgley has called 'closed instincts', humans depend on their capacity to identify and meet their full range of needs upon the conceptual resources and normative rules which constitute their local

> culture . . . Specifically culture, identity, self-realization and aesthetic
> needs interact with and complement organic needs for food and shelter in
> ways which figure less, if at all, in the ecological requirements of other
> species.
>
> (Benton, 1994: 42)

Unlike other species, human beings do not have a predetermined set of needs that must be fulfilled for them to 'flourish'. While they have in common with other animals and biological entities, 'basic' needs to meet (food, reproduction, shelter, security, sociality), there is no universal list of other needs which humans have to meet in order to flourish. Firstly, humans as a species are different from other species in having particular needs and conditions for flourishing. Secondly, some of these needs are culture-specific and cannot be generalised to all human beings.

However, this stress on the uniqueness of the human species from other species should not be read as simply re-asserting the divide between 'society' and 'culture'. We share many characteristics and needs with other animals. Thus, while we share with other social animals a need for sociality, we are different from other species in the *forms* our sociality takes, not that we are the only species in which sociality exists.

In conclusion then, a naturalistic approach to social theory, as proposed by Benton and others, can be simply stated as indicating firstly, human society and social practices by noting the commonality between the human and the natural world, and the untenability of a rigid, dualistic separation between 'society' and 'nature'. Secondly, a naturalistic perspective highlighting the centrality and constancy of the following in human life; birth, sex, death and collective subsistence, features of the 'human condition' which we share with other species, but which we fulfil and invest with normative/cultural significance in our own particular ways, ways which are not shared with other species.

In terms of the 'big issues' in human life, those aspects which have been constant features of human life around which humans have created moral codes, laws, institutions, taboos and customs, those of birth, sex, death and collective subsistence can be said to constitute a 'core' range of needs, conditions of flourishing which are central to understanding any recognisably human society. These core needs are part of life, any life whether human or nonhuman, but what marks us out is the variety of meanings we attach, the ways we make these 'natural events' or 'needs' meaningful, by marking them in ways not shared with other species. This does not deny that other species can 'grieve' for dead fellow members in ways that are intelligible to us, but even those social species which do grieve do not have anything remotely comparable to the variety of often elaborate and culturally specific ways in which humans express their grief. These cultural and morally important ways in which death, sex, birth and collective subsistence are dealt with, expressed, and the needs associated with them fulfilled are in fact what defines the 'human' and 'human society' as both a part of as well as apart from the nonhuman world.

## Conclusion

In many respects the right and left versions of how ecological and biological insights can explain (and justify) social relations is at one level another rehearsing of the old 'nature/nurture' debate, with left-wing social theories insisting on the primacy of 'nurture' and right-wing theories stressing 'nature'. The importance of these distinctions cannot be over-emphasised, since they have far-reaching consequences for the prospects for social change. If one takes the view that 'nature' (whether 'selfish genes' or the hunter-gatherer 'nature' of human beings) explains human behaviour, then the scope for changing behaviour on the basis of social change is limited. For example, if there is a genetic basis for particular types of behaviour (sexual orientation, violence, altruism and so on) then simply altering social rules or institutions will not change the genetic causes of behaviour though it might constrain it to some extent. This is because if something is 'natural', to do with 'human nature' there is very little that can be done to alter it. This type of argument is common in everyday conversation where often people put down particular types of behaviour (usually, though not always, 'bad' behaviour) to 'human nature'. 'It's only human nature isn't it?' is an all too common refrain in our daily discourse, and its significance is that it serves to both explain and close any discussion, since if it's 'human nature' then there is nothing that can be done to alter it, it just *is*. This type of argument is very common in conservative and right-wing thought, which usually starts from a negative view of human nature to defend social institutions as a way of 'managing' or perhaps lessening, but not changing or eradicating the fundamentals of human nature. However, this deterministic view is usually modified to permit some personal agency and responsibility for individual action and behaviour. Hence, a typical conservative view of crime would see it as down to the individual and having little to do with the individual's social conditions, or how they were brought up.

Against this view we have left-wing perspectives which stress social conditions, and the individual's 'social environment' including how her or she was 'socialised' or 'nurtured' as an explanation of behaviour such as crime. On this view, creating a less unequal society, one in which socio-economic opportunities were more evenly distributed, would greatly lessen criminal behaviour (though not fully eradicate it). It is from this type of thinking that the present British Prime Minister formulated his famous statement about 'Being tough on crime, and tough on the causes of crime'. Starting from a less 'negative' view of human nature and human beings, the standard left-wing position is that individuals are basically 'good' and want to be 'good' but that social and especially economic conditions prevent this from being realised, and individuals are led, encouraged or otherwise motivated to do 'bad' things. So the solution from a left-wing view is to create the social and economic conditions (or 'social environment'), by altering the present institutions of social order, to remove the socio-economic causes of criminal behaviour. Thus we come to another meaning of 'environment' which is important in social theory, environment as the 'socialising' conditions within which individuals are 'nurtured', develop their personalities, views and values. Important here is the early socialising and experience of young children, the centrality of which has been stressed by various branches of psychological theory and practice for at least a century. Part of the reason for this is that unlike most other animals, human infants are vulnerable and dependent on adults for a much longer period. This relatively long period of dependence in which the human infant is dependent on others to feed, cloth, protect it, is also the period within which the human child learns to walk, speak, think, develop the capacity for abstract thought and its sense of self. According to Dickens (1992), there is an important analogy between the 'nurturing' relationships that exist between the human infant and other humans (especially parents), and a particular 'nurturing' view of the relationship between humans and nature. For Dickens the human mind has evolved in such a way so that we are 'pre-disposed' to thinking about the natural environment as something upon which we depend. However, as pointed out in the next chapter, the environment is also something which depends on us and can be put at risk from our actions and inactions.

## Summary points

- The relationship between knowledge of the natural environment (natural sciences) and the human social world has a long and complex history.

- Sociobiological theory is an attempt to explain human social behaviour, practices and institutions based on genetics, evolutionary theory and biology. It works with a particular notion of 'human nature' in which selfishness, competitiveness and individualism are 'given' (i.e. unalterable) features of humanity.

- For critics, sociobiology is an ideology rather than science, given its close current and historical connection with right-wing and conservative political positions. This is particularly so with sociobiological attempts to establish the 'superiority' of some races over others.

- E. O. Wilson's 'biophilia' hypothesis claims that there is a genetic or evolutionary basis for human concern for the nonhuman world, which provides a sociobiological reason for conserving biological diversity.

- The concept of 'human nature' is central to social theory, and is a common reference point in arguments about how society is and ought to be. That there is no one accepted account of it illustrates its 'political' or 'ideological' character, such that one can often indicate a social theory's position on the political spectrum by its view of human nature.

- 'Progressive' social theories usually have a positive account of human nature and stress the importance of 'nurture' or the 'social environment' in explaining behaviour, while 'conservative' social arguments are usually based on a 'negative' view of an unchangeable human nature.

- A less deterministic and reductionist way of integrating biology, ecology and social theory is one which recognises that humans are both part of and apart from nature, and that humans are biologically embodied and ecologically embedded.

- We should not confuse the relevance of the natural sciences in investigating social phenomena in the human world with these sciences offering a full explanation. We cannot reduce social behaviour to the explanations of natural science.

## Further reading

A gentle (and short!) introduction to some themes in evolutionary theory is Stephen Jay Gould's *Adam's Navel* (Harmondsworth: Penguin, 1995). An excellent overview of different accounts of 'human nature' and how the latter is used in different political arguments is Christopher Berry's *Human Nature* (London: Macmillan, 1986).

Accounts of sociobiology include: Richard Dawkins' *The Selfish Gene* (2nd ed.) (Oxford: Oxford University Press, 1989); Gregory, M. *et al.* (eds), *Sociobiology and Human Nature* (San Francisco: Jossey-Bass, 1978); E.O. Wilson, *Sociobiology: The New Synthesis* (Cambridge: Belknap Press, 1980).

On the biophilia hypothesis, see E. O. Wilson, *Biophilia* (Cambridge, Mass.: Harvard University Press, 1984); Kellert, Stephen and Wilson, Edward O. (eds), *The Biophilia Hypothesis* (Washington DC: Island Press/Shearwater, 1993), and E. O. Wilson, *In Search of Nature* (London: Allen Lane, The Penguin Press, 1997).

Jared Diamond's *The Rise and Fall of the Third Chimpanzee* (London: Vintage, 1991) presents a less reactionary account of sociobiological theory.

On naturalistic social theory, see Ted Benton, *Natural Relations: Ecology, Animals and Justice* (London: Verso, 1993); Peter Dickens, *Society and Nature: Towards a Green Social Theory* (Hemel Hempstead: Harvester Wheatsheaf, 1992) and his *Reconstructing Nature: Alienation, Emancipation and the Division of Labour* (London: Routledge, 1996); Mary Midgley, *Animals and Why They Matter* (Penguin: Harmondsworth, 1983) and her *Beast and Man: The Roots of Human Nature* (Revised ed.) (London: Routledge 1995); and John Barry, *Rethinking Green Politics: Nature, Virtue and Progress* (London: Sage, 1999).

# Notes

1 URL: http: //matu1.math.auckland.ac.nz/~king/Preprints/book/diversit/restor/bph1.htm# anchor89361

2 URL: http: //matu1.math.auckland.ac.nz/~king/Preprints/book/diversit/restor/bph1.htm# anchor89361

# 9 Greening social theory

## Introduction

This chapter has a twofold purpose. The first one is to try and bring together some of the main themes and issues of the book. The second is to discuss reasons for the necessity and desirability for the 'greening' of social theory. The latter is developed through a brief discussion of the emergence and main principles of green social theory, which is then used to suggest ways in which social theory can or ought to be 'greened'. Reference will be made to *green* social, political and moral theory, which should be understood as a particular approach to the greening of social theory. That is, green social theory is only one particular way social theory can be greened, others are always possible (indeed desirable).

The call for putting the natural environment and social-environmental relations on the agenda of social and political theory is one that is in large part due to the theoretical and practical impact of green politics and philosophy over the last thirty years or so. However, as suggested in Chapters 2 and 3, we can also find some origins and antecedents of these 'green' claims which predate the rise of the modern green movement and its political, scientific and moral claims.

## Origins of green theory

Often a 'green' or 'environmental' perspective is caricatured as something that is 'counter-cultural', hippy, and out of touch with the 'realities' of modern, late twentieth-century life (see Figure 9.1). In modern Britain 'green' is most vividly associated with groups such as 'New Age Travellers' and the animal rights and anti-roads protest movement. However, the focus here is not public perceptions of 'greenies' or a sociological analysis of green groups and movements, but rather origins and principles of green social, moral and political theory.

For reasons of space, some origins of green social theory are listed below:

1 the 'romantic' and negative reaction to the industrial revolution;
2 the positive reaction to the French (democratic) Revolution;
3 a negative reaction to 'colonialism' and 'imperialism' in the nineteenth and twentieth centuries;
4 the emergence of the science of ecology;
5 growing public perception of an 'ecological crisis' in the 1960s, claims of 'Limits to Growth' in the 1970s, and the emergence of 'global environmental problems' in the 1980s and 1990s;
6 transcending the politics of 'industrialism' (organised on a left–right continuum) by a politics of 'post-industrialism' (beyond left and right);
7 increasing awareness of and moral sensitivity to our relations with the nonhuman world (from 'animal rights' to ideas that the Earth is 'sacred').

Of particular importance is the central concern of green theory and practice to overcome both the separation of 'human' from 'nature' and also the misperception of humans as above or 'superior' to nature. Green social theory can be seen as an attempt to bring humanity and the study of human society 'back down to earth'. The science of ecology played an important part in arguing that humans as a species of animal (that is, we are not just *like* animals, we *are* animals) are ecologically embedded in nature, and exist in a web-like relation to other species, rather than being at the top of some 'great chain of being'. It is crucial to note the significance of green social theory having a strong basis in the natural sciences (mainly ecology and the biological life sciences), because, as will be suggested below, this gives us a strong indication of what the 'greening' of social theory may involve.

A second and related point, is that green social theory, in transcending the culture/nature split, begins its analysis based on a view of humans as a species of natural being, which like other species has its particular species-specific characteristics, needs and modes of flourishing. Central to green social theory, unlike other forms of social theory, is a stress on the 'embodiedness' of humans.

A third issue which green social theory raises is the ways in which social-environmental relations are not only important in human society, but also *constitutive* of human society. What is meant by this is that one cannot offer a theory of society without making social-environmental interaction, and the natural contexts and dimensions of human society a central aspect of one's theory. In its attention to the naturalistic bases of human society, the green perspective is 'materialistic' in a much more fundamental way than 'Marxist materialism'. Unlike the latter, green social theory concerns itself with the external and internal natural conditions of human individual and social life, whereas the 'material base' for Marx is economic not natural. At the same time, this materialist reading of green social theory questions the 'post-materialist' character often ascribed to green politics and issues.

A fourth issue to note about green social theory is its moral claim about our relationship to the natural environment. What makes green moral theory distinctive is that it wishes to extend the 'moral community' beyond the species barrier to include our interaction with the nonhuman world as morally significant. In part, this moral concern with the nonhuman world is what gives green politics its self-professed character as 'beyond left and right'. As Lester Brown puts it, 'Both capitalists and socialists believe that humans should dominate nature. They perceive nature as a resource base to be exploited for the welfare and comfort of humans' (1989: 136).

Some of the basic principles of green social theory are listed in Box 9.1.

## Green social theory: from 'development' to 'sustainable development' and beyond

As explained in Chapter 1, modern social theory as a body of knowledge begins as the systematic study of modern or industrial society, explaining its emergence from a pre-industrial stage, and analysing its internal dynamics and processes. In this way social theory is intimately connected

**Figure 9.1 *Caricature of Greens***

*Source*: Fardell, J., *Viz Magazine*, 1992

---

**Box 9.1**

---

## Some basic principles of green social theory

Overcoming the separation between 'society' and 'environment' (which includes extending environment to include the human, built environment).

Appreciation of the *biological embodiedness* and *ecological embeddedness* of human beings and human society.

Views humans as a species of natural being, with particular species-specific needs and characteristics.

Accepts both internal and external natural limits, those relating to the particular needs and vulnerable and dependent character of 'human nature', and external, ecological scarcity in terms of finite natural resources and fixed limits of the environment to absorb human-produced wastes.

As a critical mode of social theory, green social theory criticises not just 'economic growth' but the dominant industrial model of 'development', 'modernisation' and progress.

Claims that how we treat the environment is a moral issue, and not just a 'technical' or 'economic' one. This ranges from claims that the nonhuman world has **intrinsic value**, to the idea of animal rights.

Prescriptive aspects: restructuring social, economic and political institutions to produce a more ecologically sustainable world.

'Act local, think global': ecological interconnectedness and interdependence which transcends national boundaries.

Futurity: time-frame of green social theory is expanded to include concern for future generations.

Scientific: based on ecological science (but also other natural sciences such as biology and physics).

---

with the theory and practice of the 'development' or 'modernisation' of modern societies. That is, modern social theory takes the processes of development as its object of study and aims to provide a critical analysis of *what* development consists of, *how* it occurs, *who* or what are the main features or actors of development, and *where* and *when* it occurs or has occurred. As such, the increasing concern with '**sustainable development**' within political and economic theory and practice (particularly since the 1992 Rio 'Earth Summit' conference organised by the United Nations Conference on Environment and Development), presents social theory (and the social and natural sciences more generally) with an opportunity (some might say obligation) to expand its parameters to include key aspects of this 'sustainable development' agenda (which is based upon, but not co-extensive with, the 'green' political agenda).

The essence of sustainable development is that it integrates a concern for the environment and environmental protection with obligations to present and future human generations. In terms of its most famous definition, contained in the Brundtland Report, *Our Common Future*,

> Sustainable development is development that meets the needs of the present without compromising the ability of future generations to meet their own needs. It contains within it two key concepts: – the concept of 'needs', in particular the essential needs of the world's poor, to which overriding priority should be given; and the idea of limitations imposed by the state of technology and social organisation in the environment's ability to meet present and future needs.
>
> (WCED, 1987: 43)

Sustainable development is thus development that is ecologically sustainable, that is, development that is consistent with external, natural ecological constraints and limits. Another way of looking at it has been advanced by Jacobs (1996):

> The concept of 'sustainability' is at root a simple one. It rests on the acknowledgement, long familiar in economic life, that maintaining income over time requires that the capital stock is not run down. The natural environment performs the function of capital stock for the human economy, providing essential resources and services. Economic activity is presently running down this stock. While in the short term this can generate economic wealth, in the longer term (like selling off the family silver) it reduces the capacity of the environment to provide these resources and services at all. *Sustainability is thus the goal of 'living within our environmental means'. Put another way, it implies that we should not pass the costs of present activities on to future generations.*
>
> (1996: 17: emphasis added)

The discourse (or rather discourses) of sustainability and sustainable development, acknowledge:

1  human dependence upon the natural environment, i.e. that the human economy is a subset of ecological systems;
2  the existence of external natural limits on human economic activity;

3 the detrimental effect of certain industrial activities on local and global environments;
4 the fragility of local and global environments to human collective action;
5 that one cannot talk about 'development' without also linking it to the environmental preconditions for development;
6 following on from 4, development decisions now may have environmental (and thus development) consequences for future generations and those living in other parts of the world.

In this way, green social theory, together with the emerging centrality of the theoretical and practical dimensions of sustainable development, are suggestive of why and how social theory can be 'greened'.

## Towards the greening of social theory

Green social and political theory, expressed in part, through the discourses and practices of 'sustainable development' presents at least four issues for social theory.

The first is the one at the level of the knowledge approach to the study of society. Green social theory suggests that not only should any social theory have 'social-environment' interaction as a central object of study, but that the greening of social theory requires the adoption of an explicitly multidisciplinary or interdisciplinary approach. This is because social-environmental relations and the natural contexts of human social life are so complex and involve factual as well as normative issues, which no one discipline can hope to monopolise.

The second issue concerns the temporal frame of social theory: green social theory, as expressed in its central concern with ecological sustainability and sustainable development, suggests the integration of a concern for the future and future generations. Greening social theory requires that the future be included as an explicit, rather than implicit dimension of social theory.

The third issue has to do with the brute fact that ecological problems do not respect national or cultural boundaries. Pollution problems, such as dramatically, global warming and climate change are transnational and global in scope. Thus social theory must also be transnational and global in scope and approach. It can no longer (as if it ever could) be solely concerned with a particular society independent of other countries and international processes.

The fourth issue, perhaps most contentious of all, suggests that the greening of social theory requires that social theory can no longer remain within the species boundary, that is, being solely or primarily concerned with human social relations and phenomena.

## Greening social theory: beyond 'environment' versus 'society'

The greening of social theory involves the necessary and desirable bridging of the gap between 'society' and 'nature' and also between the 'social' and 'natural' sciences. For Hayward,

> *The most distinctive green idea is that of natural relations.* These are of numerous kinds: there are natural relations of biological kinship between humans, on which familial and social relations are supervenient; between humans who are not kin, too, relations are naturally mediated, for instance in the sense that reproductive and productive activities occur in a natural medium; such activities normally involve modifying the natural environment in some way, and all humans, individually and collectively, have relations to their environment.
>
> (1996: 80; emphasis added)

This suggests that greening social theory requires a naturalistic perspective which has two main components. Firstly, a naturalistic social theory recognises the natural environmental contexts, preconditions, opportunities and constraints on human activity. Secondly, a naturalistic social theory recognises the centrality of internal human nature, of seeing humans as natural beings with particular modes of flourishing, like other natural beings.

For Ted Benton, discussed in the previous chapter, a recognition of the problems with the separation of 'society' and 'environment' and the other related distinctions between 'mind' and 'body', 'human' and 'nonhuman' (as discussed by ecofeminists in Chapter 5) requires the integration of the biologically based life sciences and the social sciences. For him, 'The task for any proposed realignment of the human social sciences with the life sciences can now be seen as providing conceptual room for organic, bodily, and environmental aspects and dimensions of human social life to be assigned their proper place' (1991: 25). This integration, as he is quick to point out, *does not* mean the reduction of one to the other. At the same time, as materialist ecofeminists point out, there needs to be a recognition of the biological reality and needs of human beings a central part of which would oblige the reorientation of social theory towards issues around reproduction and not just production (which is the narrow focus of orthodox economic theory as discussed in Chapter 6, and most of the history of pre-Enlightenment and Enlightenment political and social theory, Chapters 2 to 4).

The integration of biological and ecological insights into social theory, would produce a form of social theory which began from acknowledging human biological embodiedness and ecological embeddedness. The greening of social theory focuses on an explicit recognition of the human body and its organic needs, and fully acknowledges human limits, dependency and neediness (in part, along the lines suggested by materialist ecofeminism in Chapter 5, and non-dualistic views of the relationship between culture and nature discussed in the previous chapter). Such a social theory would see the bodily vulnerability of humans as an essential context within which to assess human mental, conceptual and non-bodily achievements and processes. Rather than seeing humans as essentially abstract, disembodied centres of reason, thinking and acting, the greening of social theory requires re-embodying the human self as a biological entity, with non-voluntary impulses, needs, instincts and feelings.

In terms of ecological embeddedness, the greening of social theory involves the acceptance of ecological limits and parameters to collective human activity. As Lee puts it, given the ecological facts of the world, and our dependent relationship with it, 'Any adequate social/moral theory must therefore address itself to these characteristics [of the world] and the character of the exchange [between humans and nature]. If it does not, whatever solution it has to offer is of no relevance or significance to our preoccupations and problems' (1989: 9).

A good example of the latter (from within sociology) is Catton and Dunlap's (1980) call for the development of a 'post-exuberant sociology' (Box 9.2). What they suggest is that the dominant paradigm within sociology (though the argument could be extended to other disciplines within the social sciences and humanities) is a 'human exceptionalist' one based on the dominant 'Western worldview' which they claim is out of keeping with the ecological reality, context and limits of human societies.

# Box 9.2

## Catton and Dunlap's call for a 'post-exuberant' sociology

### Assumptions of the dominant 'human exceptionalist' paradigm

'1. About the *nature of human beings* – people are fundamentally different from all other creatures on earth, over which they have dominion.

2. About the *nature of social causation* – people can determine their own destinies, can choose their goals and learn whatever is necessary to achieve them.

3. About the *context of human society* – the world is vast and provides unlimited opportunities for humans.

4. About the *constraints on human society* – the history of human society is one of progress, there is a solution to every problem, and progress need never cease.'

(Catton and Dunlap, 1980: 34)

### Assumptions of the 'new ecological paradigm'

'1. Humans have exceptional characteristics but remain one among many in an interdependent global ecosystem.

2. Human affairs are influenced not only by social and cultural factors but also by the complex interactions in the web of nature, so that human actions can have many unintended consequences.

3. Humans are dependent on a finite biophysical environment which sets physical and biological limits to human affairs.

4. Although the inventiveness of humans and the power derived therefrom may seem for a while to extend carrying capacity limits, ecological laws cannot be repealed.'

(Catton and Dunlap, 1980: 34)

# Ecology: connecting the natural and social sciences?

Ecology or the 'ecological paradigm' has, for some, since its emergence in the last century, promised a 'unified science of nature and society'. Ecology and its aim of a 'naturalistic' account of the human condition, while originating as an empirical natural science dealing with the relationship between species and their environments, has also become a form of social and moral theory. According to Vincent, 'ecology has had moral and religious import for humanity' (1992: 213), noting however that 'there is a tangled and uneasy relationship between those who perceive ecology as an established science and those who mesh the scientific findings with strong doses of normative theory' (1992: 209).

However, the science of ecology (dealing with facts and the way the natural world is) has tended to go hand in hand with normative claims (dealing with values and how the world should be, and how we ought to treat and use nature), and has found it difficult to maintain a strict and lasting separation between 'facts' and 'values'. Eroding this strict distinction has placed ecology in a unique position as a 'science', as a form of knowledge which seems to bridge the natural and social sciences. This can be seen in how 'environmental studies' or 'environmental management' programmes in universities necessarily have to straddle the social and natural sciences, reflecting the interdisciplinary character of the forms of knowledge appropriate to articulating social-environmental relations. This of course goes against Habermas's ideas discussed in Chapter 4, in which only natural science and an instrumental/technical relationship to the environment was deemed to be the most 'fruitful' approach to take.

## A unified science of humanity and nature

Greening social theory seems to take us in the direction of challenging strict disciplinary boundaries, not just within the social sciences, but also between the social and the natural sciences. According to Dickens,

> I would argue that a new paradigm is now being forced by environmental issues and related matters such as human health, animal welfare and the application of new reproductive technologies to the human species. *Such a paradigm rejects the distinctions between, for example, the life*

*sciences, the physical sciences and the social sciences.* It is nevertheless a combination of these apparently alternative ways of viewing the social and natural worlds, within a coherent epistemological framework.

(1992: 2; emphasis added)

This idea of a unified and integrated science of humanity and nature is something that can be found in the thought of Karl Marx. For him, 'The idea of *one* basis for life and another for *science* is from the very outset a lie . . . Natural science will in time subsume the science of man just as the science of man will subsume natural science: there will be *one* science' (Marx, 1975: 355).

While the greening of social theory may not (and perhaps should not) move in the direction of the emergence of a unified, homogeneous and singular 'science', it is clear that overcoming the culture/human versus nature/environment dichotomy also takes us in the direction of weakening and loosening the boundaries between different forms of human knowledge. As Hayward (1995) notes, given that the economy is the most important and visible aspect of human society which interacts with the natural environment, it is clear that the greening of social theory will necessarily have to have an economic dimension. As he points out, 'If a unified theory of economics and ecology is to be possible, it will neither hypostatize (sic) an opposition between economy and ecology nor posit a straightforward identity of the two' (1995: 116). Rather, in the spirit of returning to the older tradition political economy, the greening of social theory will, in part, be a theory of political ecology within which is located the study of political economy.

## Greening social theory 1: future generations

One of the central claims of green social theory is its concern to extend the temporal dimension of social theory to include a sense of futurity. This reorientation of social theory towards the future is at the heart of the idea of sustainable development, with its explicit recognition that present decisions about how we treat and use the environment cannot be taken without considering the likely impact of these decisions on the type of environment that we leave to future generations, and thus the effect the latter may have on the welfare of the future or on the ability of the future to meet its needs. Thus the greening of social theory requires lengthening the temporal frame of social theory.

A sense of the future dimension involved is the Native American Indian saying that 'We don't inherit the earth from our parents, but borrow it from our children'. How far into the future is of course an open and important question, the answer to which will have important implications for society. However, the key issue is that greening social theory implies extending its temporal range into the future, in particular making the likely ecological impacts of present courses of action an explicit object of analysis.

## Greening social theory 2: beyond the nation-state and globalisation

The greening of social theory requires the adoption of an international and global perspective, since social-environmental relations are not always contained within a specific geographical area and the effects of social intervention in the environment do not always respect national boundaries. In this way the greening of social theory requires that social theory be self-consciously international and global in its outlook and analysis. As Yearley points out, 'environmental dangers pose supranational problems; these need solutions to which national governments are not well suited' (1991: 45).

Another important dimension of the greening of social theory is the process of **globalisation**. While social theory has always had an international dimension (though often the various links between societies were assumed rather than made an explicit object of study), the main focus of social theory was on the internal dynamics of society. With the advent of global and transnational environmental problems, highlighting the ecological interconnections between geographically separated parts of the earth, the study of the origins and effects of and alternatives to these various transnational environmental problems necessarily requires that one must look beyond (and sometimes below) the nation-state. The interconnectedness of the world's economic and ecological systems means that what happens within a society increasingly has its origins in processes outside that society. This means that one must look not just beyond the boundaries of the nation-state but also at other international actors other than the nation-state. These include: powerful global non-government actors such as transnational corporations (TNCs); weaker, but still significant environmental non-government organisations (NGOs); and transnational political and economic institutions such as the United Nations and the World Bank.

The work of Giddens and Beck discussed earlier in Chapters 4 and 7, is concerned with the way in which theorising about the environment necessitates a global or international perspective. Put another way, social theorising about the environment is a particular aspect of social theorising about globalisation. This link between globalisation and the environment can be seen in Paterson's view that, 'if globalisation is environmentally problematic . . . the politics of those concerned with environmental problems lies in resisting globalising forces, such as multinationals, banks and governments, in their attempts to negotiate new international regimes on the environment' (1996: 403–4).

The greening of social theory here requires incorporating the ideas of green thinkers and activists who argue that the global political and economic system maintains an exploitative relationship between the affluent Northern countries (the so-called 'developed' world) and the poor Southern countries (the so-called 'developing' world). Here the greening of social theory suggests analysing the cultural, political, economic and ecological effects of globalisation within the context of North–South relations. For example, while global environmental threats such as global warming and climate change affect everyone on the planet, this does not mean that humanity as a whole is in the same boat. Firstly, it is not 'humanity' as a whole that is to blame for global environmental problems; those with economic and political power, mainly 'developed' industrialised nations, are mostly to blame for causing environmental problems, by, for example, being responsible for the vast bulk of global carbon pollution from the use of fossil fuels. Secondly, environmental problems do not have the same effects on everyone. So for example, affluent countries and groups are better able to protect themselves from environmental problems than poor countries and groups. As Seabrook notes,

> *Globalisation is not an organic growth*, but a carefully wrought ideological project . . . If the perpetuation of privilege is to be the guiding force of the world, let it be identified as such . . . If the abuse of the resource-base of the earth and the intensifying exploitation of its people is the supreme civilisational pursuit of the culture of globalisation, no matter what the consequences, why shrink from saying so?
>
> (1998: 27; emphasis added)

Rejecting the claim that globalisation is not something natural or organic, that is, something which is both 'given', 'inevitable' and beyond human control, and also somehow 'good', has recently played a major part in critiques of parties and governments such as Tony Blair's New Labour administration. According to a recent edition of *Marxism Today* (November 1998), the New Labour government views globalisation as 'a force of nature' (Jacques, 1998), an inevitable and irresistible economic process that the power of the nation-state cannot affect. Against such a 'natural' and inevitable phenomenon all we can do is adapt ourselves, our economies and societies to this current phase of the 'evolution' of the world economy. The result of seeing globalisation as a force of nature can explain why New Labour has much in common with the previous Conservative administration with regard to the economy. As Stuart Hall points out, '*Since globalisation is a fact of life* to which There Is No Alternative, and national governments cannot hope to regulate or impose any order on its processes or effects, New Labour has accordingly largely withdrawn from the active management of the economy' (1998: 11; emphasis added). In this way viewing or describing something as a force of nature, a 'fact of life' or 'natural' has tremendous ideological power in prescribing particular forms of action, or in the case of New Labour, inaction.

Since environmental degradation is caused by poverty and socio-economic inequality, stopping environmental destruction requires alleviating poverty, which in turn requires the creation of a fairer global economic system. The latter, according to Gray (1998), in turn requires regulating the global economy. While many greens would go beyond Gray, and argue for the restructuring or dismantling of the global economy, his argument for the political regulation of the global economy, for ecological and social reasons, is something that fits with the main thrust of green social theory in respect to globalisation. According to Gray,

> What is beyond doubt is that organising the world economy as a single global free market promotes instability. It forces workers to bear the costs of new technologies and unrestricted free trade. It contains no means whereby activities which endanger the global ecological balance can be curbed. Organising the world economy as a universal free market is, in effect, staking the planet's future on the supposition that these vast dangers will be resolved as an unintended consequence of the unfettered pursuit of profits. It is hard to think of a more reckless danger.
>
> (1998: 32)

## Greening social theory 3: beyond the species barrier

According to Brian Baxter, 'it is now intellectually unacceptable to develop political theories in which the sole focus of concern is human well-being and values, ignoring the issues which greens have pushed to the fore concerning the well-being of other species, and the biosphere in general' (1996: 68). Baxter makes this claim on the basis that there are compelling moral arguments (from within green theory) which mean that it is illegitimate not to extend moral considerability and concern beyond humans to include (at least parts of) the nonhuman environment (animals, plants, ecosystems). This normative approach to including the natural environment and its interests in human social theory has much to commend it, even thought it goes against many settled assumptions about our attitudes towards and treatment of the nonhuman world.

However, one can also advance reasons why the nonhuman world should be part of the agenda of social theory on the more factual grounds that social-environmental relations are constitutive of human societies. That is, we cannot fully grasp or understand any human society without also understanding the types and meanings attached to the ways in which society views, values, treats and uses its natural environment. Of particular importance here is the relationship between domesticated animals and humans, where it is true to say that human societies and cultures in which relations with domesticated animals were absent would be unintelligible to us in ways that societies which have these relations would not present the same problem.

At the same time, there are those like Milton (1996) whose ideas suggest that social theory, at least that part of social theory which is based on reflecting on culture, cannot remain concerned with human culture. As she puts it, 'As we learn more about both human and nonhuman animals, *it becomes increasingly difficult to sustain the view that culture is uniquely human*' (1996: 64; emphasis added). In this effort she is joined by some recent moral philosophers who hold that morality, in the sense of acting or behaving in accordance with moral principles, like 'culture', is not something that is unique to humans (see Singer, 1994: part B).

Social theory can be (indeed it must be) *human-based* but does not have to be *human-centred*. That is, while social theory must begin from an analysis of human society, it does not necessarily have to be exclusively concerned with human affairs, interests and events. For example, given both the co-evolutionary history of humans and animals, as well as the various material ways in which human societies use and have relationships with animals, social theory can reorientate its aims and objects of study to include these nonhumans as fellow members of 'society'.

# Conclusion

It is rather paradoxical that what started out as an examination of the ways in which social theory has viewed, used and abused the external environment should at various points (in different ways, and for different reasons) oblige us to return inwards towards an examination of 'human nature' and its place in social theory. This was chartered in numerous ways in the last chapter where we saw how the introduction of environment to social theory (naturalising social theory) also leads us in the direction of integrating biological concerns, while in Chapter 3 we saw how arguments based on the 'state of nature' also included assumptions about 'human nature'.

Andrew Brennan puts this nicely when he notes that, 'in order to discover what sort of human life is valuable we must first consider what kind of thing a human being is. Although there is, in my view, no complete answer to this question, we can . . . grasp one important aspect of human nature by reflecting on what are essentially ecological considerations' (1988: xiii). In this way one could say that by its very nature (excuse the pun) when we use the word 'environment' we almost automatically move in the direction of talking about 'external nature' and from there it is difficult to prevent this discussion spilling over into debates about 'internal nature'.

We begin by analysing the natural world around us (that which environs us) and our relations to and with it, and find that this cannot be done without reflecting on 'us' as embodied beings, a particular species of animal evolved from social primates, and our own 'nature'. And the ultimate paradox of humanity is that the 'human condition' (which is also at the same time our 'natural condition') is one in which humanity is both a part of and apart from the environment.

## Summary points

- The main principles of green theory are a rejection of the separation of 'humanity' and 'environment'; a stress on the biological embodiedness and ecological embeddedness of humans; viewing social-environmental relations as not only important in human society, but also constitutive of human society; and a claim that social-environment relations are of moral concern.

- The ideas behind 'sustainable development' can be used as a way to explore the main ways in which green social theory can and does contribute to the 'greening' of social theory in general.

- The greening of social theory involves overcoming a strict separation between 'environment' and 'society', and stresses the ecological embeddedness and biologically embodiedness of human beings and human society.

- Ecology and the 'ecological paradigm' has been used to suggest the integration of the natural and social sciences, which for some may lead to a 'unified science of humanity and nature'.

- While not perhaps going so far as a unified science of humanity and nature, the greening of social theory requires the adoption of a more multi- or interdisciplinary approach.

- Three dimensions of the greening of social theory include: integrating a concern for future generations, extending social theory beyond the nation-state and focusing on globalisation; and finally, extending social theory beyond the species barrier, such that while social theory must be human-based it does not necessarily have to be human-centred.

- The greening of social theory and the social theorising about the environment can be seen as focusing on the paradox of the 'human condition', in which humans are both a part of yet apart from the natural environment.

## Further reading

On green social and political theory: good introductory overviews include Stephen Young's *The Politics of the Environment* (Manchester: Baseline Books, 1992), and Robert Garner's *Environmental Politics* (Hemel Hempstead: Harvester Wheatsheaf, 1995).

More detailed accounts are my own *Rethinking Green Politics: Nature, Virtue and Progress* (London: Sage, 1999), Andy Dobson's *Green Political Thought: An Introduction* (London: Routledge, 2nd edn 1995), Luke Martell's *Nature and Society: An Introduction* (Cambridge: Polity, 1994), Robert Goodin's *Green Political Theory* (Cambridge: Polity, 1992), Robyn Eckersley's *Environmentalism and Political Theory: Towards an Ecocentric Approach* (London: UCL Press, 1992), Tim Hayward's *Ecological Thought: An Introduction* (Cambridge: Polity, 1995).

On the greening of social theory see William Catton and Riley Dunlap's 'A New Ecological paradigm for a Post-Exuberant Sociology', *The American Behavioral Scientist*, 24: 1 (1980); and for book-length treatments, see the excellent works: Peter Dickens, *Society and Nature: Towards a Green Social Theory* (Hemel Hempstead: Harvester Wheatsheaf, 1992); Ted Benton, *Natural Relations:*

*Ecology, Animal Rights and Social Justice* (London: Verso, 1993); and Keekok Lee, *Social Philosophy and Ecological Scarcity* (London: Routledge, 1989). On 're-embodying' social theory see Chris Shilling, *The Body and Social Theory* (London: Sage, 1993).

On green social theory and globalisation, see *The New Internationalist*, 'Globalization: Pealing Back the Layers', November 1997, and 'Currencies of Desire', October 1998; Tim Lang and Colin Hines, *The New Protectionism* (London: Earthscan, 1993); and ch. 5 of Kay Milton's *Environmentalism and Cultural Theory* (London: Routledge, 1996).

 **Glossary**

**acid rain**  rain, snow or mist in which water droplets have combined with acidic gases such as sulphur dioxide, usually as a result of the burning of fossil fuels, such as coal, gas or petroleum.

**anthropocentrism**  thinking or acting that is predominantly concerned with humans. It can take 'weaker' or 'stronger' forms. The strong version holds that human interests or purposes are the only issue in making moral judgements, while the weaker version holds that while human interests are important, they are not the only ones to be considered.

**biodiversity**  shorthand for biological diversity, referring to the variety and quantity of species of plant and animal life.

**classical liberalism**  an early form of liberal theory the central principles of which were a 'laissez-faire' or free market view of the economy, a rejection of state interference in economic and social relations, all with the aim of defending a particularly individualistic view of human liberty. Modern versions of classical liberal themes can be found in contemporary libertarianism.

**ecocentrism**  thinking or acting that is predominantly concerned with both humans and nonhumans.

**ecology**  has a few different meanings. It can mean the science of ecology, the branch of biology concerned with the relations of living things (mainly animals and plants) and their environments, or it can mean the 'ecological paradigm' that is a summary of green ideas which contain both factual and normative claims.

**the Enlightenment**  sometimes called 'modernity'. Denotes the radical series of changes in European thought and action which occurred towards the end of the eighteenth century.

**fact/value distinction** refers to an important distinction within social theory between statements about the way things are, or descriptions about the way things are (facts), and statements about how things ought to be or normative judgement about things (values).

**fossil fuels** are non-renewable forms of energy such as coal, oil, gas, timber and peat.

**genetic engineering and biotechnology** human manipulation of genetic information and material to create new forms of organisms.

**globalisation** denotes the series of cultural, political, economic, environmental and social changes and developments which are turning the world into a single market, creating webs of relations bringing different parts of the world closer together into one unified system.

**instrumental value** something has instrumental value if its value (or worth) is given by reference to some other purpose or entity.

**intrinsic value** something which has value in itself as opposed to being valuable only for some other purpose or entity.

**libertarianism** modern heir of the classical liberal view of society which argues for minimal state interference in the economy, coupled with a belief in the efficiency and liberty-enhancing effects of the unfettered free market, all aimed at realising a particular negative form of individual freedom as 'freedom from' state and social interference.

**mode of production** in Marxist political economy this refers to the particular mode or ways in which a society produces its means of subsistence (it includes the level and type of technology, property relations, and the division of labour in a society), each society, according to the Marxist theory of historical materialism will have a different mode of production.

**modernisation** is a highly complex and contentious term, which is sometimes used as another word for 'development'. It describes a particular social development path, largely based on the historical experience of European societies. A modernising society displays a series of changes in its economy, politics and culture, such as a move from an agricultural, rural economy to an industrial, urban one, a shift from non-democratic forms of government to (liberal) democratic forms, and the secularisation of values and culture and the decline of traditional or religious values.

**natural order**  has its roots in historical accounts of the relationship between humans and nature, referring to the 'given' (often God-given) and 'proper' organisation of the world including human society.

**NGOs/Non-Governmental Organisations**  are voluntary groups or organisations which operate independently of national and international government agencies and institutions. They are often international in focus and often lobby governments while monitoring government activity in a variety of fields including environmental and economic policy.

**political economy**  a term predating modern economics as a discipline but which deals with much the same subject matter: how to deal with the 'economic problem' of limited resources and infinite human wants, and the efficient allocation of resources with competing uses. Where political economy differs from modern economics is in its explicit recognition of the political context of economic activity, and the links between politics and economics.

**social construction**  refers to a particular approach within social theory which holds that either there is no 'objective' reality for humans which has not been constructed via language, or that there is always a social or discursive dimension to human reality.

**steady-state economy**  a term first used by ecological economist Herman Daly to describe the following: a constant population level, a constant capital stock, and minimising energy and matter 'throughput' in the economy.

**sustainability**  has a variety of meanings, but the main thrust of the concept conveys a sense of something continuing (lasting, enduring sustaining) into the future. Usually taken to mean ecological sustainability in the sense of 'maximum yield', a sustainable use of resources which does not undermine regeneration rates.

**sustainable development**  in the words of the Brundtland Commission, sustainable development means the ability of the present generation to meet its needs without undermining the ability of the future to meet its own needs.

**technocentric**  used to describe approaches to social and environmental problems as 'technical' matters which can be solved by technological or technical means, usually the application of scientific or expert knowledge and/or some technological innovation.

**utilitarianism** the moral theory in which the 'greatest happiness' of the 'greatest number', the balance of pleasure over pain, or utility over disutility is the criterion by which one ought to judge different courses of action.

# Internet resources and sites

There a thousands of environmental and green sites, some are academic, some are government-related and many are maintained by environmental groups, green parties, movements and activists.

Some good places to start your search are:

1 **Fundamentally Green**  http://www.barnsdle.demon.co.uk/pol/fundi.html
2 **EnviroLink**  http://www.envirolink.org
3 **GreenNet Home Page**  http://www.gn.apc.org
4 **Political Science Resources**  http://www.psr.keele.ac.uk/

The latter also has links to a wide variety of sites on political and social theory

On social theory also see:  http://www.geocities.com/~sociorealm/socialth2.htm

## Environmental groups/movements

1 **Green Parties of North America:**  http://www.rahul.net/greens/
2 **Green Party of England and Wales:**  http://www.greenparty.org.uk/
3 **The European Federation of Green Parties:**  http://www.dru.nl/maatschappij/politiek/groenen/europe/europe.htm
4 **Worldwide Fund for Nature:**  http://www.panda.org/
5 **Earthaction:**  http://www.oneworld.org/earthaction/
6 **Greenpeace:**  http://www.greenpeace.org/
7 **Eco- The Campaign for Political Ecology:**  http://www.gn.apc.org/eco/
8 **The Land Is Ours:**  http:www.envirolink.org/orgs/tlio/
9 **Centre for Alternative Technology:** http://www.foe.co.uk/CAT/

10 **Earthfirst!:** http://www.hrc.wmin.ac.uk/campaigns/ef/earthfirst.html
11 **Rainforest Action Network:** http://www.ran.org
12 **Environmental Protest in Britain:** http://www.keele.ac.uk/depts/po/pol/courses/m307/britprot.htm
13 **The New Economics Foundation:** http://sosig.ac.uk/New Economics/newecon.html

## Journals, magazines and bibliographical sites

1 **Bibliography on Biodiversity:** http://osu.orst.edu/dept/ag_resrc_econ/ biodiv/biblio.html
2 **Electronic Green Journal:** http://www.lib.uidaho.edu:70/docs/egj.html
3 **Red Pepper Magazine:** http://www.netlink.co.uk/users/editoria/
4 **Environmental Newsletter E-Zine:** http://www.geocities.com/Eureka/Plaza/1697/newsletter.html
5 **EcoSocialist Review:** http://www.dsausa.org/dsa/ESR/index.html
6 **National Library for the Environment from CNIE:** http://www.cnie.org/nle/
7 **International Society for Environmental Ethics:** http://www.cep.unt.edu/ISEE/html
8 **Green Politics Newsletter:** http://www.keele.ac.uk/depts/po/pol/green/march98.htm
9 **New Internationalist:** http://www.oneworld.org/ni/

## Gender and environment, ecofeminism

1 **Gender and Urban Planning:** http://www-rcf.usc.edu/~harwood/fem&plan.htm
2 **Women in Natural Resources:** http://www.ets.uidaho.edu/winr/index.
3 **Women and Environments:** http://www.web.net/~weed
4 **Ecofeminism:** http://www2.infoseek.com/Titles?qt=ecofeminism

## Government and academic sites on the environment and environmental issues

1 **European Environment Agency:** http://www.eea.dk
2 **Centre for Social and Economic Research on the Global Environment:** http://www.uea.ac.uk/env/cserge/noframe.htm
3 **Department of Environment, Transport and the Regions (UK):** http://www.detr.gov.uk/itwp/index.htm
4 **Southampton Library Environmental Resources:** http://www.southampton.liunet.edu/library/environ.htm
5 **Global Environmental Change Site:** http://www.sussex.ac.uk/Units/gec/
6 **National Centre for Sustainability (US):** http://www.islandnet.com/~ncsf/ncsf/homemenu.htm
7 **Sustainable Development:** http://www.ulb.ac.be/ceese/sustul.htm
8 **Institue for Bioregional Studies:** http://www.cycor.ca/IBS/
9 **Systematic Work on Environmental Ethics:** http://www.cep.unt.edu/theo.html
10 **Centre for Study of Social Movements (Cantebury):** http://snipe.ukc.ac.uk/sociology/polsoc.html
11 **US Global Change Research Program:** http://www.usgcrp.gov
12 **Biodiversity and Ecosystem Network (BENE):** http://straylight.tamu.edu/bene/bene.html
13 **Centre for World Indigenous Studies:** http://www.halcyon.com/FWDP/cwisinfo.html

# Bibliography

Alper, J. (1978) 'Ethical and Social Implications', in Gregory, M. *et al.* (eds).

Andermatt Conley, V. (1997) *Ecopolitics: The Environment in Poststructuralist Thought*, London: Routledge.

Arnold, D. (1996) *The Problem of Nature: Environment, Culture and European Expansion*, Oxford: Blackwell.

Bandarage, A. (1997) *Women, Population and Global Crisis: A Political-Economic Analysis*, London: Zed Books.

Barnes, M. (ed.) (1994) *An Ecology of the Spirit*, Lanham, Maryland: University Press of America.

Barry, J. (1990) 'Limits to Growth', unpublished MA dissertation.

Barry, J. (1993) 'Deep Ecology and the Undermining of Green Politics', in J. Holder *et al.* (eds).

Barry, J. (1994) 'Beyond the Shallow and the Deep: Green Politics, Philosophy and Praxis', *Environmental Politics*, 3: 3.

Barry, J. (1995a) ' Towards a Theory of the Green State', in S. Elworthy *et al.* (eds).

Barry, J. (1995b) 'Nature in Question: What is the Question?', *Environmental Politics*, 4: 1.

Barry, J. (1995c) 'Deep Ecology, Socialism and Human 'Being in the World: A Part of Yet Apart from Nature', *Capitalism, Nature, Socialism*, 6: 3.

Barry, J. (1996a), 'Democracy, Judgement and Sustainability', in B. Doherty and M. Geus (eds).

Barry, J. (1998a) 'Green Political Theory', in A. Lent (ed.).

Barry, J. (1998b) 'Social Policy and Social Movements: Ecology and Social Policy' in N. Ellison and C. Pierson (eds).

Barry, J. (1998c) 'Marxism and Ecology', in A. Gamble *et al.* (eds).

Barry, J. (1999) *Rethinking Green Politics: Nature, Virtue and Progress*, London: Sage.

Baxter, B. (1996) 'Must Political Theory Now Be Green?', in I. Hampshire-Monk and J. Stanyer (eds).

Baxter, B. (1999) *Ecologism: A Defence*, Edinburgh: Edinburgh University Press.

Beck, U. (1992a) *Risk Society: Towards a New Modernity*, London: Sage.

Beck, U. (1992b) 'From Industrial Society to Risk Society: Questions of Survival, Social Structure and Ecological Enlightenment', *Theory, Culture and Society*, 9: 1.

Beck, U. (1995a) *Ecological Enlightenment: Essays on the Politics of the Risk Society*, Princeton, NJ: Humanities Press International.

Beck, U. (1995b) *Ecological Politics in an Age of Risk*, Cambridge: Polity Press.

Bentham, J. (1823/1970) *The Principles of Morals and Legislation*, Darien, CT: Hafner Publishing.

Benton, T. (1989) 'Marxism and Natural Limits: An Ecological Critique and Reconstruction', *New Left Review*, 178.

Benton, T. (1991) 'Biology and Social Science: Why the Return of the Repressed Should be Given a (Cautious) Welcome', *Sociology*, 25: 1.

Benton, T. (1993) *Natural Relations: Ecology, Animals and Social Justice*, London: Verso.

Benton, T. (1994) 'Biology and Social Theory in the Environmental Debate', in M. Redclift, and T. Benton (eds).

Benton, T. and Reclift, M. (1994) 'Introduction' in M. Redclift and T. Benton (eds) *Social Theory and the Global Environment*, London: Routledge.

Berry, C. (1986) *Human Nature*, London: Macmillan.

Bookchin, M. (1986) *The Modern Crisis*, Philadelphia: New Society Publishers.

Boulding, K. (1966) 'The Economics of the Coming Spaceship Earth', in H. Jarrett (ed.), *Environmental Quality in a Growing Economy*. Baltimore: Johns Hopkins Press.

Bramwell, A. (1989) *Ecology in the 20th Century: An Introduction*. London and New Haven: Yale University Press.

Brennan, A. (1988) *Thinking about Nature: An Investigation of Nature, Value and Ecology*, London: Routledge.

Brown, L. (1989) *Envisioning a Sustainable Society: Learning Our Way Out*, New York: State University of New York Press.

Cahn, M. and O'Brien, R. (1996) *Thinking About the Environment: Readings on Politics, Property and the Physical World*, New York and London: M.E. Sharpe.

Callicott, J.B. (1994) *Earth's Insights: A Multicultural Survey of Ecological Ethics from the Mediterranean Basin to the Australian Outback*, Berkeley: University of California Press.

Carson, R. (1962) *Silent Spring*, Greenwich, CT: Fawcett Premier.

Cassell, P. (ed.) (1993) *The Giddens Reader*, London: Macmillan.

Castree, N. (forthcoming), 'Nature', in K.J. Johnston *et al.* (eds) *The Dictionary of Human Geography*, 4th edn, Oxford: Blackwell.

Catton, W. and Dunlap, R. (1980) 'A New Ecological Paradigm for a Post-Exhuberant Sociology, *American Behavioral Scientist*, 24: 1.

*The Children's Dictionary* (1969), London: The Grolier Society.

Clayre, A. (ed.) (1977) *Nature and Industrialization*, Oxford: Oxford University Press.

Collard, A. (1988) *Rape of the Wild: Man's Violence against Animals and the Earth*, Indianapolis: Indiana University Press.

Condorcet, M. (1995) 'The Future Progress of the Human Mind', in I. Kramnick (ed.).

Cooper, D. (1992) 'The Idea of Environment' in D. Cooper and J. Palmer (eds).

Cooper, D. and Palmer, J. (eds) (1992) *The Environment in Question: Ethics and Global Issues*, London: Routledge.

Croall, S. and Rankin, W. (1981) *Ecology for Beginners*, New York: Pantheon Books.

Croll, E. and Parkin, D. (1992) *Bush Base, Forest Farm: Culture, Environment and Development*, London: Routledge.

Daly, H. (ed.) (1973) *Toward a Steady-State Economy*. San Francisco: W.H. Freeman.

de Geus, M. (1999) *Ecological Utopias: Envisioning the Sustainable Society*, Utrecht: International Books.

de-Shalit, A. (1997) 'Is Liberalism Environment-Friendly?', in R. Gottlieb (ed.).

Diamond, C. (1978) 'Eating Meat and Eating People', *Philosophy*, 53.

Diamond, J. (1991) *The Rise and Fall of the Third Chimpanzee*, London: Vintage.

Dickens, P. (1992) *Society and Nature: Towards a Green Social Theory*, London: Harvester Wheatsheaf.

Dobson, A. (1993) 'Critical Theory and Green Politics', in A. Dobson and P. Lucardie (eds).

Dobson, A. (1995) *Green Political Thought*, 2nd edn, London: Routledge.

Dobson, A. and Lucardie, P. (eds) (1993) *The Politics of Nature: Explorations in Green Political Theory*, London: Routledge.

Docherty, T. (1993) 'Postmodernism: An Introduction', in T. Docherty (ed.).

Docherty, T. (ed.) (1993) *Postmodernism: A Reader*, Hemel Hempstead: Harvester Wheatsheaf.

Doherty, B. and de Geus, M. (eds) (1996) *Democracy and Green Political Theory: Sustainability, Justice and Citizenship*, London: Routledge.

Doyle, T. and McEachern, D. (1998) *Environment and Politics*, London: Routledge.

Dryzek, J. (1987) *Rational Ecology: Environment and Political Economy*. Oxford: Basil Blackwell.

Dryzek, J. (1990) 'Green Reason: Communicative Ethics for the Biosphere', *Environmental Ethics*, 12: 3.

Eckersley, R. (1992) *Environmentalism and Political Theory: Toward an Ecocentric Approach*, London: UCL Press.

Eder, K. (1996) *The Social Construction of Nature: A Sociology of Ecological Enlightenment*, London: Sage.

Eklund, R. and Herbert, R. (1975) *A History of Economic Theory and Method*, New York: McGraw-Hill.

Ekins, P. (ed.) (1986) *The Living Economy: A New Economics in the Making*, London: Greenprint.

Ekins, P. (ed.) (1992) *A New World Order: Grassroots Movements for Global Change*, London: Routledge.

Elliot, D. (1997) *Energy, Society and Environment*, London: Routledge.

Ellison, N. and Pierson, C. (eds) (1998) *Developments in British Social Policy*, London: Macmillan.

Engel, J. and Engel, J. (eds) (1990) *The Ethics of Environmental Development*, London: Belhaven Press.

Evans, J. (1993) 'Ecofeminism and the Politics of the Gendered Self', in A. Dobson and P. Lucardie (eds).

Foster, J. (1997) 'Introduction', in J. Foster (ed.).

Foster, J. (ed.) (1997) *Valuing Nature? Economics, Ethics and Environment*, London: Routledge.

Freud, S. (1950), '"Civilized" Sexual Morality and Modern Nervousness', *Collected Papers*, vol. II, London: Hogarth Press.

Gandy, M. (1996) 'Crumbling Land: The Postmodernity Debate and the Analysis of Environmental Problems', *Progress in Human Geography*, 20: 1.

Gare, A. (1995) *Postmodernism and the Environmental Crisis*, London: Routledge.

Giarini, O. (1980) *Dialogue on Wealth and Welfare*, Pergamon Press.

Giddens, A. (1985) *The Nation-State and Violence*, Cambridge: Polity.

Giddens, A. (1990) *The Consequences of Modernity*, Cambridge: Polity.

Giddens, A. (1991) *Modernity and Self-Identity: Self and Society in the Late Modern Age*, Cambridge: Polity.

Giddens, A. (1994) *Beyond Left and Right: The Future of Radical Politics*, Cambridge: Polity.

Giddens, A. *et al.* (1994) 'Introduction', in Giddens, A. *et al.* (eds), *The Polity Reader in Social Theory*, Cambridge: Polity.

Glacken, C. (1967) *Traces on the Rhodian Shore*, Berkeley: University of California Press.

Goldblatt, D. (1996) *Social Theory and the Environment*, Cambridge: Polity.

Goodin, R. (1992) *Green Political Theory*, Cambridge: Polity.

Gottlieb, R. (ed.) (1996) *This Sacred Earth: Religion, Nature, Environment*, London: Routledge.

Gottlieb, R. (ed.) (1997) *The Ecological Community: Environmental Challenges for Philosophy, Politics and Morality*, London: Routledge.

Gould, P. (1988) *Early Green Politics*, Brighton: Harvester Wheatsheaf.

Gray, J. (1998) 'Globalisation: The Dark Side', *New Statesman*, 13 March.

Gregory, M. (1978) 'Epilogue' in M. Gregory *et al.* (eds).

Gregory, M. *et al.* (eds), (1978) *Sociobiology and Human Nature*, San Franscico: Jossey-Bass.

Guha, R. and Martinez-Alier, J. (1997) *Varieties of Environmentalism: Essays North and South*, London: Zed Books.

Habermas, J. (1975) *Legitimation Crisis*, Boston: Beacon Press.

Habermas, J. (1981) 'New Social Movements', *Telos*, 49.

Habermas, J (1982), 'A Reply to My Critics', in J. Thompson and D. Held (eds).

Hampsher-Monk, I. and Stanyer, J (eds) (1996) *Contemporary Political Studies 1996*, vols 1–3, Belfast: Political Studies Association.

Hannigan, J. (1995) *Environmental Sociology: A Social Constructionist Perspective*, London: Routledge.

Haraway, D. (1995) 'Otherwordly Conversations, Terrain Topics, Local Terms', in V. Shiva and I. Moser (eds) *Biopolitics: A Feminist and Ecological Reader in Biotechnology*, London: Zed Books.

Haught, T. (1994) 'Religion and the Origins of the Environmental Crisis', in M. Barnes (ed.).

Hayward, T. (1995) *Ecological Thought: An Introduction*, Cambridge: Polity.

Hayward, T. (1996) 'What Is Green Political Theory?', in I. Hampshire-Monk and J. Stanyer (eds).

Heilbroner, R. (1967) *The Worldy Philosophers*, New York: Simon & Schuster.

Henderson, H. *et al.* (1986) 'Indicators of No Real Meaning', in P. Ekins (ed.)

Hodgson, G. (1997) 'Economics, Environmental Policy and the Transcendence of Utilitarianism', in J. Foster (ed.).

Holland, A. (1997) 'Substitutability: Or, Why Strong Sustainability is Weak and Absurdly Strong Sustainability Is not Absurd', in J. Foster (ed.).

Horkheimer, M. and Adorno, T. (1973) *Dialectic of Enlightenment*, London: Allan Lane.

Hughes, J. D. (1994) 'Ecology in Ancient Mesopotamia', in D. Wall (ed.).

*Hutchinson Dictionary of Ideas* (1994), Oxford: Helicon.

Inglehart, R. (1977) *The Silent Revolution: Changing Values and Political Styles among Western Publics*, Princeton: Princeton University Press.

Ingold, T. (1992) 'Culture and the Perception of the Environment', in E. Croll and D. Parkin (eds) *Bush Base, Forest Farm: Culture, Environment and Development*, London: Routledge.

Jacobs, M. (1994) 'The Limits of Neoclassicalism: Towards an Institutional Environmental Economics', in M. Redclift and T. Benton (eds), *Social Theory and the Environment*. London: Routledge.

Jacobs, M. (1996) *The Politics of the Real World*. London: Earthscan.

Jacques, M. (1998) 'Editorial', *Marxism Today*, November/December.

Jameson, F. (1991) *Postmodernism; or, the Cultural Logic of Late Capitalism*, London: Verso.

Keat, R. (1994) 'Citizens, Consumers and the Environment: Reflections on *The Economy of the Earth*', *Environmental Values*, 3: 4.

Kellert, S. and Wilson, E.O. (eds) (1993), *The Biophilia Hypothesis*, Washington DC: Island Press/Shearwater.

Kidner, D. (1994) 'Why Psychology Is Mute about the Environmental Crisis', *Environmental Ethics*, 16: 4.

Kinsley, David (1996) 'Christianity as Ecologically Harmful', in R. Gottlieb (ed.).

Kohák, E. (1984) *The Embers and the Stars: A Philosophical Inquiry into the Moral Sense of Nature*, Chicago and London: University of Chicago Press.

Kramnick, I. (ed.) (1995) *The Portable Enlightenment Reader*, London: Penguin.

Lash, S., Szersynski, B. and Wynne, B. (eds) (1996) *Risk, Environment and Modernity: Towards a New Ecology*, London: Sage.

Latouche, S. (1993) *In the Wake of the Affluent Society*, London: Zed Books.

Layder, D. (1994) *Understanding Social Theory*, London: Sage.

Lee, K. (1989) *Social Philosophy and Ecological Scarcity*, London: Routledge.

Leeson, S. (1979) 'Philosophic Implications of the Ecological Crisis: The Authoritarian Challenge to Liberalism', *Polity*, 11: 303–18.

Lent, A. (ed.) (1998) *New Political Thought: An Introduction*, London: Macmillan.

Levidow, L. (1995) 'Whose Ethics for Agricultural Biotechnology?', in V. Shiva and I. Moses (eds).

Levidow, L. and Tait, J. (1995) 'The Greening of Biotechnology: GMOs and Environment-Friendly Products', in V. Shiva and I. Moses (eds).

Lewontin, R. (1982) 'Organism and Environment', in H. Plotkin (ed.).

Lyotard, J. (1984) *The Postmodern Condition: A Report on Knowledge*, Minneappolis: University of Minnosota Press.

McKibben, B. (1989) *The Death of Nature*, New York: Random House.

Macpherson, C.B. (1973) *Democratic Theory*, Oxford: Clarendon Press.

Marcuse, H. (1955) *Eros and Civilization: A Philosophical Inquiry into Freud*, New York: Vintage Books.

Marcuse, H. (1964) *One Dimensional Man*, Boston: Beacon Press.

Marcuse, H. (1972) *Counterrevolution and Revolt*, London: Penguin.

Marcuse, H. (1992) 'Ecology and the Critique of Modern Society', *Capitalism, Nature, Socialism*, 3: 3.

Marshall, P. (1995) *Nature's Web: Rethinking Our Place on Earth*, London: Cassell.

Martinez-Alier, J. (1995) 'Political Ecology, Distributional Conflicts, and Economic Incommensurability', *New Left Review*, 211.

Martin-Brown, J. (1992) 'Women in the Ecological Mainstream', *International Journal*, XLVII: 4.

Marx, K. and Engels, F. (1848/1967) *The Communist Manifesto*, Harmondsworth: Penguin.

Marx, K. (1975) 'Economic and Philosophical Manuscripts', in L. Colletti (ed.), *Karl Marx: Early Writings*, Harmondsworth: Penguin

Masters, R. (1991), 'Jean-Jacques Rousseau' in D. Miller *et al.* (eds).

May, T. (1996) *Situating Social Theory*, Buckingham: Open University Press.

Mellor, M. (1992a) *Breaking the Boundaries: Towards a Feminist, Green Socialism*, London: Virago.

Mellor, M. (1992b) 'Green Politics: Ecofeminist, Ecofeminine or Ecomasculine?', *Environmental Politics*, 1: 2.

Mellor, M. (1997) *Feminism and Ecology*, Cambridge: Polity Press.

Merchant, C. (1990) *The Death of Nature*, New York: Harper & Row.

Mill, J.S. (1848/1900), *Principles of Political Economy*, London: Longmans, Green and Co.

Mill, J.S. (1977) 'On Nature', in A. Clayre (ed.).

Miller, C. (1988) *Jefferson and Nature: An Interpretation*, Baltimore: Johns Hopkins University Press.

Miller, D. (1991) 'Peter Kropotkin', in D. Miller *et al.* (eds).

Miller, D. *et al.* (eds) (1991) *The Blackwell Encyclopaedia of Political Thought*, Oxford: Blackwell.

Milton, K. (1996) *Environmentalism and Cultural Theory: Exploring the Role of Anthropology in Environmental Discourse*, London: Routledge.

Mirowski, P. (1994) 'Doing What Comes Naturally: Four Metanarratives on What Metaphors Are For', in P. Mirowski (ed.).

Mirowski, P. (ed.) (1994) *Natural Images in Economic Thought: Markets Read in Tooth and Claw*, Cambridge: Cambridge University Press.

Mulberg, J. (1995) *Social Limits to Economic Theory*, London: Routledge.

Nash, R. (1967) *Wilderness and the Ameican Mind*, New Haven and London: Yale University Press.

Nisbet, R. (1982) *The Social Philosophers*, New York: Washington Square Press.

Northcott, M. (1996) *The Environment and Christian Ethics*, Cambridge: Cambridge University Press.

O'Brien, R. and Cahn, M. (1996) 'Thinking About the Environment: What's Theory Got to Do with It?', in M. Cahn and R. O'Brien.

O'Connor, M. T. (ed.) (1995) *Is Capitalism Sustainable?: Political Economy and the Politics of Ecology*, New York and London: Guildford Press.

O'Neill, J. (1993) *Ecology, Policy and Politics: Human Well-Being and the Natural World*, London: Routledge.

O'Riordan, T. (1981) *Environmentalism*, 2nd revised edn, London: Pion.

O'Riordan, T. and Jordan, A. (1994) 'The Precautionary Principle in Contemporary Environmental Politics', *Environmental Values*, 4: 3.

Oelschlaeger, M. (ed.) (1992) *The Wilderness Condition: Essays on Environment and Civilization*, Washington and Covelo, CA: Island Press.

Offe, C. (1984), *Contradictions of the Welfare State*, London: Hutchinson.

Outhwaite, W. (ed.) *The Habermas Reader*, Oxford: Polity Press.

Parsons, H. (1977) *Marx and Engels on Ecology*, Westport, CT: Greenwood Press.

Passmore, J. (1980) *Man's Responsibility for Nature*, 2nd edn, London: Duckworth.

Paterson, M. (1996) 'UNCED in the Context of Globalisation', *New Political Economy*, 1: 3.

Pearce, D. (1992) 'Green Economics', *Environmental Values*, 1: 1.

Pearce, D., Markandya, A. and Barbier, E. (1989) *Blueprint for a Green Economy*, London: Earthscan.

Pepper, D. (1984) *The Roots of Modern Environmentalism*, London: Croom Helm.

Pepper, D. (1996) *Modern Environmentalism: An Introduction*, London: Routledge.

Pietilä, H. (1990) 'The Daughters of Earth: Women's Culture as a Basis for Sustainable Development', in J. Engel and J. Engel (eds).

Plant, J. (1989) 'Toward a New World Order: An Introduction', in J. Plant (ed.).

Plant, J. (ed.) (1989) *Healing the Wounds: The Promise of Ecofeminism*, Philadelphia: New Society Publishers.

Plotkin, H. (ed.) (1982) *Learning, Development and Culture*, Chichester: Wiley.

Plumwood, V. (1993) *Feminism and the Mastery of Nature*, London: Routledge.

Polanyi, K. (1947) *The Great Transformation: The Political and Economic Origins of Our Time*. Boston: Beacon Press.

Porteous, J. D. (1997) *Environmental Aesthetics: Ideas, Politics and Planning*, London: Routledge.

Porter, R. (1994) 'The Enlightenment', in *The Hutchinson Dictionary of Ideas*, Oxford: Helicon.

Purdue, D. (1995) 'Hegemonic Trips: World Trade, Intellectual Property Rights and Biodiversity', *Environmental Politics* 4: 1.

Redclift, M. and Benton, T. (eds) (1994) *Social Theory and the Global Environment*, London: Routledge.

Rennie-Short, J. (1991) *Imagined Country: Society, Culture and Environment*, London: Routledge.

Robertson, G., Mash, M., Tickner, L., Bird, J., Curtis, B., and Putnam, T. (eds) (1996) *FutureNatural: Nature, Science, Culture*, London: Routledge.

Rose, S., Lewontin, R. and Kamin, L. (1984) *Not in Our Genes*, Harmondsworth: Penguin.

Rousseau, J.J. (1995), 'A Critique of Progress', in I. Kramnick (ed.), *The Portable Enlightenment Reader*, London: Penguin.

Sallah, A. (1992) 'The Ecofeminist /Deep Ecology Debate: A Reply to Patriarchal Reason', *Environmental Ethics*, 14: 3.

Sallah, A. (1995) 'Nature, Woman, Labor, Capital: Living the Deepest Contradiction', in M. O'Connor (ed.).

Salleh, A. (1997) *Ecofeminism as Politics: Nature, Marx and the Postmodern*, London: Zed Books.

Seabrook, J. (1998) 'A Global Market for All', *New Statesman*, 26 June.

Shiva, V. (1988) *Staying Alive: Women, Ecology and Development*, London: Zed Books.

Shiva, V. (1992) 'Overview', *Ms Magazine*, 11: 6, 3.

Shiva, V. and Moses, I. (eds) (1995) *Biopolitics: A Feminist and Ecological Reader on Biotechnology*, London: Zed Books.

Singer, P. (1990) *Animal Liberation*, 2nd edn, London: Jonathon Cape.

Singer, P. (ed.) (1994) *Ethics*, Oxford: Oxford University Press.

Smith, N. (1996) 'The Production of Nature', in G. Robertson *et al.* (eds).

Soper, K. (1995) *What Is Nature?*, Oxford: Blackwell.

Spencer, H. (1884/1982) *The Man versus the State*, Indianapolis: Liberty Press.

Spretnak, C. (1996) 'Beyond Humanism, Modernity, and Patriarchy', in R. Gottlieb (ed.).

Stephens, P. (1996) 'Plural Pluralisms: Towards a More Liberal Green Political Theory', in I. Hampsher-Monk and J. Stanyer (eds).

Sumner, W. (1992) *On Liberty, Society and Politics: The Essential Essays of William Graham Sumner*, Indianapolis: Liberty Press.

Swartz, Daniel (1996) 'Jews, Jewish Texts, and Nature: A Brief History', in R. Gottlieb (ed.).

Thompson, J. and Held, D. (eds) (1982) *Habermas: Critical Debates*, London: Macmillan.

Thomas, K. (1983) *Man and the Natural World: Changing Attitudes in England 1500–1800*, Harmondsworth: Penguin.

Tocqueville, A. (1956) *Democracy in America* (ed. R. Heffner), New York: Mentor Books.

van den Bergh, J. (1996) *Ecological Economics and Sustainable Development: Theory, Methods and Application*, Cheltenham: Edward Elgar.

Vincent. A. (1992) 'Ecologism', in his *Modern Political Ideologies*, Oxford: Blackwell.

Vogel, S. (1997) 'Habermas and the Ethics of Nature', in R. Gottlieb (ed.).

Wall, D. (1994) *Green History: A Reader in Environmental Literature, Philosophy and Politics*, London: Routledge.

Warren, K. (1987) 'Feminism and Ecology: Making Connections', *Environmental Ethics*, 9: 1.

White, L. (1967) 'The Historical Roots of Our Ecologic Crisis', *Science*, 155.

Whitebrook, J. (1996) 'The Problem of Nature in Habermas', in D. Macauley (ed.), *Minding Nature: The Philosophers of Ecology*, New York and London: Guildford Press.

Williams, R. (1988), *Keywords: A Vocabulary of Culture and Society*, London: Fontana.

Wilson, E. O. (1978) 'What Is Sociobiology?', in M. Gregory *et al.* (eds).

Wilson, E.O. (1993) 'Biophilia and the Conservation Ethic', in S. Kellert and E.O. Wilson (eds).

Wilson, E. O. (1997) *In Search of Nature*, London: Allen Lane, the Penguin Press.

Wissenburg, M. (1998) *The Free and the Green Society: Green Liberalism*, London: UCL Press.

Woodgate, G. and Redclift, M. (1998) 'From a "Sociology of Nature" to Environmental Sociology: Beyond Social Construction', *Environmental Values*, 7: 1.

World Commission on Environment and Development (1987), *Our Common Future*, Oxford: Oxford University Press.

Worster, D. (1994) *Nature's Economy: A History of Ecological Ideas*, 2nd edn, Cambridge: Cambridge University Press.

Yearley, S. (1991) *The Green Case: A Sociology of Environmental Issues, Arguments and Politics*, London: HarperCollins.

Zimmerman, M. (1992) 'The Blessing of Otherness: Wilderness and the Human Condition', in M. Oelschlaeger (ed.).

Zimmerman, M. (1994) *Contesting Earth's Future: Radical Ecology and Postmodernity*, Berkeley, University of California Press.

# Index